Advances in Asian Human-Environmental Research

Aims and Scope

The series aims at fostering the discussion on the complex relationships between physical landscapes, natural resources, and their modification by human land use in various environments of Asia. It is widely acknowledged that human-environment interactions become increasingly important in area studies and development research, taking into account regional differences as well as bio-physical, socioeconomic and cultural particularities.

The book series seeks to explore theoretic and conceptual reflection on dynamic human-environment systems applying advanced methodology and innovative research perspectives. The main themes of the series cover urban and rural landscapes in Asia. Examples include topics such as land and forest degradation, glaciers in Asia, mountain environments, dams in Asia, medical geography, vulnerability and mitigation strategies, natural hazards and risk management concepts, environmental change, impacts studies and consequences for local communities. The relevant themes of the series are mainly focused on geographical research perspectives of area studies, however there is scope for interdisciplinary contributions.

More information about this series at http://www.springer.com/series/8560

Christoph Bergmann

The Himalayan Border Region

Trade, Identity and Mobility in Kumaon, India

Springer

Christoph Bergmann
South Asia Institute
Heidelberg University
Heidelberg, Germany

ISSN 1879-7180 ISSN 1879-7199 (electronic)
Advances in Asian Human-Environmental Research
ISBN 978-3-319-29705-7 ISBN 978-3-319-29707-1 (eBook)
DOI 10.1007/978-3-319-29707-1

Library of Congress Control Number: 2016937711

Printed on acid-free paper

This Springer imprint is published by Springer Nature
The registered company is Springer International Publishing AG Switzerland

Foreword

The scholarly debate on borderlands and transboundary environments represents an emerging research field in the social sciences. Since the early nineteenth century, the mountain regions of South and Central Asia, including the Hindu Kush, Karakoram, Himalaya, Pamir, and the Tibetan Plateau, have been key arenas for imperial expansion, territorial restructuring, and border-making. However, in-depth analyses of these sociopolitical processes, particularly with regard to their localized manifestations, are still extremely rare. Christoph Bergmann meets a part of this critical knowledge and research gap in the border triangle between India, the Tibet Autonomous Region (TAR, People's Republic of China), and Nepal, where contestations over political boundaries have engendered multifarious challenges as well as opportunities for local mountain communities. His deep-seated analysis on the Bhotiyas, former trans-Himalayan traders residing in several high valleys of the Indian Kumaon Himalaya, is the outcome of a doctoral dissertation titled "Living in a High Mountain Border Region: Politics and Poetics of Central Himalayan Pastoralism." The thesis was part of an interdisciplinary research project where social anthropologists and geographers collaborated in an endeavor to better understand pastoral livelihoods in the Himalayan border region.

This book provides an insightful discussion on the trade, identity, and mobility of the Bhotiyas who, over long periods of time, have become adept in positioning themselves favorably within a dynamic field of actor and interest constellations. The analysis draws from extensive archival work and long-term ethnographic research in the Darma valley, located in the easternmost part of Kumaon. Christoph Bergmann offers a rare glimpse into how the Bhotiyas' trans-Himalayan trade evolved in an imperial environment under British colonial rule, how their tribal identities were contested and forged in postcolonial times, and how their pastoral mobility was refashioned in the face of marketization and political reforms during India's economic liberalization after 1991. In all the empirical chapters, the author efficiently interweaves the narratives and day-to-day activities of the Bhotiyas with the wider geopolitical contexts – ranging from the British Empire's territorial expansions in the nineteenth and early twentieth centuries to the ramifications of independent India's defeat in its border war with China in 1962 and from India's national policies and programs for tribal development to the power struggles associated with the creation of the new mountain state of Uttarakhand in 2000.

The empirically and theoretically rich book is exemplary in three main ways: first, it demonstrates a deep ethnographic knowledge of a fascinating and little known borderland community; second, it fuses insights and concepts from two disciplines – anthropology and geography; and, third, it makes an original and historically grounded contribution to the emerging scholarly discourse on political agency in high mountain environments. Altogether Christoph Bergmann opens new interdisciplinary pathways for human-environmental research, at the juncture of politics and poetics.

I highly recommend reading this book to anyone interested in understanding how and why the Himalayan border region has evolved as an "agentive" site of action for a multitude of different actors.

Heidelberg, Germany
December 2015

Marcus Nüsser

Preface

This book is the outcome of my Ph.D. research, which I conducted at the South Asia Institute, Heidelberg University. My field research in India would never have been possible without the warm hospitality of the Darmani. Special thanks go to all the residents of the villages Son and Dugtu for their trust and affection. I was to feel at home during my prolonged field research and was assisted in all possible ways. The people to whom I owe a profound debt are too many to mention, but I must name Jasvant S. Sonal and his family for "adopting" me as their son, Jevan S. Sipal for introducing me to many villages at the beginning of my field research, Daram S. Sonal and his family for always providing a safe harbor in Dharchula, Dabla S. Phirmal for sharing his profound knowledge and for his tremendous efforts in translating materials from the Darma language into Hindi, Moti R. Sonal for his amazing helpfulness and friendship, and Laxman S. Kutiyal for assisting me in expanding my personal network in Dharchula and beyond. I am further deeply indebted to Syam C. Kharkwal as well as to Lalit Joshi and his family for their support.

I gratefully acknowledge the generous financial support of the State Graduate Sponsorship Program of Baden-Württemberg (doctoral scholarship, 2008–2010) and the German Research Foundation (NU 102/10-1 and 10-2, 2010–2013). I had the immense privilege of joining the History Department at the Kumaon University Nainital, India, as a research fellow between 2008 and 2010. I would further like to thank Girija Pande for his invaluable support and advice; my colleagues Eva Ambos, Sandra Bärnreuther, Paul Roden, and Christian Strümpell for their constant encouragement and close reading of earlier chapter drafts; Jamuna S. Bhattarai, Jagadish Bhattarai, and Bimal Diwadi for their assistance in translating fieldwork materials into English; Bernhard Dickoré for sharing his unmatched botanical expertise; and my teachers and mentors, William S. Sax and Marcus Nüsser, who inspired me with their passion for the Himalayan border region, provided a most valuable intellectual base for my work, and fostered an interdisciplinary cooperation that is exceptional.

Earlier chapter drafts were presented at the Kumaon University Nainital (India, 2008); a joint conference of the Collaborative Research Center 619, the Cluster of Excellence "Asia and Europe in a Global Context" and the DFG-project NU-102/10-1 in Heidelberg (Germany, 2010); the second conference of the Asian Borderlands Research Network in Chiang Mai (Thailand, 2010); the WIN-conference at the Heidelberg Academy of Sciences (Germany, 2011); the meeting of the German Society for Anthropology in Vienna (Austria, 2011); the "Ritual Journeys" conference of the Collaborative Research Center 619 in Heidelberg (Germany, 2012); the third conference of the Asian Borderlands Research Network (Singapore, 2012); the workshop "Living in Asian Borderlands: An Appreciation of Resources" at the University of Tübingen (Germany, 2014); and the SSAI graduate workshop "Crossing borders in South Asia" at SOAS, University of London (UK, 2014). I am very grateful to the organizers of these events for providing me with an opportunity to present my research and to the participants who responded to my work and helped to refine my thoughts.

Very special words of thanks go to my family: my wife Mónica, who accompanied me to the field in 2008–2009 and supported me during the long and sometimes difficult journey of my Ph.D. project; my daughter Julia, who joined our travels in the Himalaya with so much joy; and my parents, who have always supported my efforts. I will always miss the company of my colleague and dear friend Martin Gerwin, with whom I had the fortune to cover a long distance of the research process but who could unfortunately not witness its final outcome.

Heidelberg, Germany Christoph Bergmann

Contents

1 Introduction 1
1.1 Border Studies, or: Appreciating the Centrality of the Himalayan Border Region 5
1.2 Who Are the Bhotiyas? 7
1.3 Geographies of Trade, Identity and Mobility in Kumaon 10
1.4 Processes of Negotiation in a Transforming Socio-spatial Landscape 13
1.5 Methodological Approach 15
1.6 How the Book Is Organized 16
References 18

2 Trans-Himalayan Trade in an Imperial Environment 23
2.1 Approaching Territoriality and Sovereignty in the Himalayan Border Region 26
2.2 Trans-Himalayan Trade Within Pre-colonial Power Constellations 29
2.3 Moorcroft's Journey, or: The Discovery of Black Pudding 38
2.4 Securing Traders' Loyalty in a Non-Regulation Province 43
2.5 Advancing the Territorial Structuring of Imperial Space 48
2.6 Claiming Territorial Sovereignty and Negotiating for the Formal Rule of Law 54
2.7 Coda 60
References 64

3 Tribal Identities and Scalar Politics in Postcolonial India 75
3.1 The Notion of 'Tribe' 81
3.2 Ethnic Identities and Scalar Politics 84
3.3 Old Lady Jasuli 88
3.4 Producing *Bhot* as a Scale Effect in Independent India 92
3.5 Reproducing Independent India as a Scale Effect in *Bhot* 100

3.6 Negotiating 'Bhotiya' as a Scalar Expression 107
3.7 Coda 116
References 120

4 Facets of Pastoral Mobility in a Market-Oriented Border Region 127
4.1 The Bhotiyas' Seasonal Migration in the Context of Emergent Policy Initiatives 134
4.2 Rituals as an Embodied Practice of Pastoral Mobility 142
4.2.1 Establishing Fields of Involved Activity 145
4.2.2 The *Alam*-Exchange 149
4.2.3 Celebrating Harvest 155
4.3 Dealing with Death Along Routes of Travel 160
4.4 Coda 177
References 180

5 Epilogue 185
References 188

Index 191

Abbreviations

BADP	Border Area Development Program
BAMCEF	Backward and Minority Community Employees' Federation
BBUC	Border Bhotiya Uplift Committee
BJP	Bharatiya Janata Party
BRO	Border Roads Organisation
BSTSC	Bhotiya Scheduled Tribe Struggle Committee
BYWA	Border Youth Welfare Organization
CoI	Constitution of India
DWDP	Decentralized Watershed Development Project
GoI	Government of India
GoU	Government of Uttaranchal/Uttarakhand
GoUP	Government of Uttar Pradesh
INR	Indian rupees
ITBP	Indo-Tibetan Border Police
KBPF	Kumaon Bhotiya Peoples' Federation
KMVN	Kumaon Mandal Vikas Nigam Ltd.
LAC	Line of Actual Control
NEFA	North-East Frontier Agency
NGO	Non-governmental organization
OBCs	Other Backward Classes
RWA	Rang Welfare Association
SC	Scheduled Caste
SCI	Supreme Court of India
SDM	Subdivisional District Magistrate
ST	Scheduled Tribe
TAC	Tribes Advisory Council
TAR	Tibet Autonomous Region of the People's Republic of China
TCBL	Traditional Customary Boundary Line
THED	Theory of Himalayan Environmental Degradation

TSP	Tribal Sub-Plan
UHC	Uttaranchal/Uttarakhand High Court
UP	Uttar Pradesh
USD	U.S. dollar
USSS	Uttarakhand Sanyukta Sangarsh Samiti (Uttarakhand United Struggle Association)
USTG	Uttaranchal State Tribal Organization
UTDP	Uttarakhand Tourism Development Plan

Chapter 1
Introduction

Abstract This chapter introduces the so-called Bhotiyas, former trans-Himalayan traders and a Scheduled Tribe of India. The Bhotiyas reside in several high valleys of the Kumaon Himalaya, which borders the Tibet Autonomous Region (China) to the north and Nepal to the east. Recent advances in the growing field of border studies are discussed, which establishes an analytical framework that allows, first, to appreciate the centrality of this Himalayan border region within the entangled imperial and national histories of adjacent states and, second, to explore the agency of its marginalized residents in dealing proactively with shifting forms of territorial boundedness. On this basis I foreground the competing forms of power that have unevenly reshaped Kumaon's geographies of trade, identity and mobility since colonial times. I then provide some insights into how the Bhotiyas have harnessed their endogenous capacities and capitalized on external forces of change whilst negotiating their positionality within a transforming socio-spatial landscape. The chapter closes with an outline of my methodological approach and the overall structure of the book.

Keywords Trade • Identity • Mobility • Agency • Border studies • Kumaon Himalaya • India

It was in June 2009, with most of the heavy work of ploughing and sowing complete, that the residents of Son, a high altitude and seasonally inhabited village in the northeastern part of India's Kumaon Himalaya, began their preparations for a special celebration (Photo 1.1). Everybody was abuzz with excitement. Smoke rose through the stone tiles of the communal kitchens, and the smell of food wafting and shouts of greetings resounding through the alleys as urban dwelling relatives were welcomed home in preparation for this rare event that takes place only every fourth to sixth year. As with most of the villagers, I too was looking forward to the initiation of the ceremony with a ritual that had not been carried out for decades, and that only the old were familiar with. The ritual is called *sun dum*. *Sun* translates as 'village' and *dum* as 'to chase away'. As the naming suggests, it is performed to chase someone or something away from the spheres of village life – fortunately not the researcher, but some evil spirit (*syino*). For that purpose all residents – high- and low-caste, young and old, male and female – met at an open space near the southern

C. Bergmann, *The Himalayan Border Region*, Advances in Asian Human-Environmental Research, DOI 10.1007/978-3-319-29707-1_1

Photo 1.1 Son village in the Darma valley (Photograph © Christoph Bergmann, August 30, 2009)

and lower entrance of the village after sunset. Residents understand that all kinds of negative influences and forces, such as diseases, political turmoil or evil spirits, come from this direction.

After everybody had armed him- or herself with a machete or axe there occurred a moment of confusion, since many residents had not performed the ritual for quite a long time or even never throughout their whole life. The situation clarified, however, when the ritual specialist of the village arrived after some delay. He had a goat on a lead and was carrying a small and somewhat scary-looking figure, called *miju*. It was made from the dough of three different types of roasted grain, namely two local varieties of buckwheat and one variety of wheat. Its orifices were marked with dried red chilly peppers and black peppercorns. After having placed the *miju* on the ground, the ritual specialist instructed all high-caste residents to form a circle around it. He then handed over the goat to a low-caste person who stood somewhat apart from the main group with his fellows. Without further ado that man began orbiting the crowd, whilst wielding the branches of a nettle-like shrub over people's heads. The ritual specialist (RS), with whom the low-caste person (LP) responsively exclaimed the following phrases, chased after him:

(LP) All villagers have come together.
(RS) Throw it away!
(LP) From today you stop putting us in trouble!
From today you stop afflicting us!

Today you receive your part of chilly and salt!
We give you the *miju*!
Today you receive your colored *daja* [ribbon]
We give you black lentils,
and we also give you meat.
From today you stop afflicting us!
(RS) Throw it away!
(LP) All families of the village have given their share!
(RS) Throw it away!
(LP) You evil spirit go!

When the ritual specialist had eventually caught up with the low-caste person, he, without hesitation, beheaded the goat with his machete right on the spot. All participants then touched the carcass with their blades in order to get some blood on it. In the meantime the ritual specialist prepared a cardboard box in which he put the *miju*, a piece of meat from the animal just killed, some salt, black lentils and a small bundle of colored ribbons (*daja*). The low-caste person was then ordered by him to throw the box down a deep gorge into the main river running through the valley. The remaining low-caste villagers were expected to take the animal carcass and share its meat once everyone else had left the site. This, however, did not take long. Families rushed home and entered their houses after having placed their weaponry at the doorposts with the bloody blades exposed. Gathered around the hearth inside the kitchen of the house, the family members had some food whilst awaiting the low-caste person to visit them. As soon as that man announced his arrival at the front door of the house, people rushed to the entrance stairs and sprinkled him with water. This should prevent the malevolent influences with which he had come in contact to gain entrance. Only then was the man invited to take a seat next to the kitchen door, where he was offered some food and liquor.

The Darma valley, of which Son village forms part, is located in the Indian state of Uttarakhand, which borders the Tibet Autonomous Region (China) to the north and Nepal to the east (Fig. 1.1). Its inhabitants, the Darmani, consist of two main social strata, namely high-caste Rajputs and low-caste Harijans. The Government of India notified the former as a Scheduled Tribe (ST), the so-called Bhotiyas, in 1967.[1] This status grants these former trans-Himalayan traders access to specific

[1]The term ‘Rajput’ literally translates as ‘son (*putra*) of a king (*raja*)’. In the western and central Himalaya members of the warrior or Kshatriya class (*varna*) are widely referred to as Rajputs. However, as Sax (2002: 95) argues, the term is problematic because “different people have different opinions about who is and who is not a ‘true’ Rajput”. While it is in general applied to high-caste groups, people traditionally distinguished between immigrant (Thuljat) and indigenous (Khasa) Rajputs, with the latter ranking lower than the former. The Thuljat Rajputs, to which some Bhotiyas claim to belong, trace their ancestry to high-caste immigrants who entered the area in the late medieval period and became its ruling class (Berreman 1963; Sanwal 1976). The lowest castes, which had formerly been referred to as ‘untouchables’, are nowadays most often addressed as Harijans (Sax 2009). The term was coined by Gandhi and literally translates as ‘child of god’. Alternatively, the politically more engaged term ‘Dalit’, which literally means ‘oppressed person’, is used. In the Darma valley residents indicate their status by using different middle names, namely ‘Singh’ for high-caste and ‘Ram’ for low-caste. The surname is the same and derived from the village name, such as Sonal for the inhabitants of Son village.

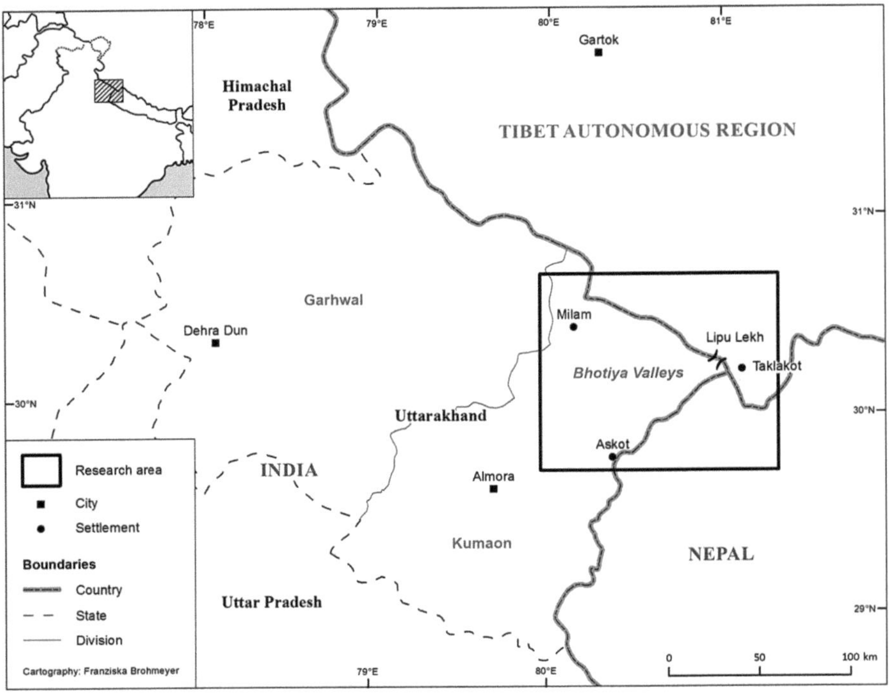

Fig. 1.1 Research area in the Himalayan border region of Kumaon (Bergmann 2016: 90)

entitlements with regard to social welfare, education and employment. Such measures of affirmative action are meant to ensure their autonomy and decrease their vulnerability as an autochthonous national minority. Valley dwellers practice 'combined mountain agriculture' (Ehlers and Kreutzmann 2000), i.e. a mix of animal husbandry, crop cultivation and forest and pasture use in different altitudinal belts. Throughout the year a migratory household shifts between two main locations, namely a high altitude summer settlement, such as Son village, and a winter settlement in the lower parts of the valley.

One might be tempted to describe ritual performances such as *sun dum* as part of an ancient, presumably stable and somewhat peculiar tradition. Such an appraisal would fit the still popular image that pastoralist tribes in remote mountain areas are 'primitive' in terms of their livelihood strategies, ritual practices, and religious beliefs. However, as the present study argues, such performances rather hint toward a culturally informed and highly creative dimension of human agency. This agency is used proactively, such as for establishing trade relations, asserting an identity or maintaining a high level of mobility in the course of ongoing state and market reforms. It can also, simultaneously, serve to reinforce inequalities along the lines of caste, class or gender. In any case, it forms part and parcel of the manner in which borderland populations, like the Darmani, engage "with powerful external normalizing factors" (Michaud 2011: 219): from the destructive powers of an evil spirit to the ramifications of shifting state policies and territorial restructurings; and from increasingly pervasive markets to broader processes of environmental, technological and social change.

Whilst the agentive use of ritual is not a characteristic giving the Bhotiyas a unique and exceptional status, their agency has evolved within and draws its force from a specific field of socio-ecological production. Based on a shared positionality in the border triangle between India, Tibet and Nepal these people could explore opportunities and experiment with solutions that are not available to communities in other parts of this mountain range. In the following I discuss recent advances in the growing field of border studies, which establishes an analytical framework that allows, first, to appreciate the centrality of this Himalayan border region within the entangled imperial and national histories of adjacent states and, second, to explore the agency of its marginalized residents in dealing proactively with shifting forms of territorial boundedness.

1.1 Border Studies, or: Appreciating the Centrality of the Himalayan Border Region

While anthropologists have investigated social boundaries for some time (Barth 1969), national state borders have only recently attracted wider scholarly attention (Alvarez 1995; Baud and van Schendel 1997; Paasi 2009; Johnson et al. 2011; Wilson and Donnan 2012).[2] In the course of this steadily growing interest, it has been argued that borders and border regions are a 'crucial key' for understanding the cultural predicaments of the twenty-first century (Wendl and Rösler 1999: 1), and even, as Rumford (2006: 155) puts it, "to understand the nature of the social". By relying on a perspective from the periphery, borderland scholarship shows that historical processes of territorial and economic restructuring are influenced not only by state policies but are also shaped by the strategies of resident communities (Anderson and O'Dowd 1999). While states rely heavily on the broadening of a bureaucratic machinery and all kinds of infrastructure facilities to foster a lasting influence at their peripheries (Vandergeest and Peluso 1995; Walker 1999), these efforts intermingle with and are contested by borderlanders' aspirations, giving rise to competing interests and power struggles (Sturgeon 2004).[3]

[2] Specific case studies on South Asian border regions are given, for instance, in van Schendel (2005), Warikoo (2009), Zutshi (2010), Gerwin and Bergmann (2012), Jones (2012), Gellner (2013), Harris (2013), Baghel and Nüsser (2015), and Shneiderman (2015).

[3] The concepts 'frontier', 'border', 'borderland', and 'boundary' contain slightly different connotations. According to Wendl and Rösler (1999: 3) a crucial etymological distinction between the words 'frontier' and 'border' is that the former term implicates "an absolute, anthropocentric body-based standpoint" while the latter "denotes a bird's-eye view, with the observer not bodily involved". Frontiers are often associated with the territorial expansion of nations and empires into areas that are considered 'empty'. F.J. Turner (1894) has famously coined this understanding with reference to the westward expansion of the American frontier. Borders, by way of contrast, can be

A multitude of powerful actors have produced, contested and transformed the socio-ecological fabric of the Himalayan border region in Kumaon. Between the eighteenth and nineteenth century the region marked an effective barrier to both British and Chinese imperial expansion. In the twentieth century this old imperial frontier was turned into a national state border under military surveillance, a process that consolidated after the Sino-Indian border war in 1962. Since then the destiny of the rising global powers of India and China has become irrevocably entwined with their inability "to share the same mountains" – the Himalaya – and to finally settle remaining territorial disputes along the common border (Malone 2011: 152). When India began to liberalize its market in the early 1990s (Kohli 2006), commercial interests increasingly affected geopolitical imperatives and advanced new patterns of regional restructuring (Gerwin and Bergmann 2012). While rapid economic growth as well as the rising international significance of both countries attracted much public and academic attention, related and equally compelling issues remained largely ignored:

- How have communities such as the Bhotiyas dealt with imperial forms of power and authority in order to realize their trans-Himalayan trade under British colonial rule?
- How have they performed various scales of identity and political power to assert who they are or want to be in postcolonial times?
- How did they re-fashion their pastoral mobility, including associated rituals, following India's implementation of liberal economic reforms since the early 1990s?

In the present study I seek to answer these questions based on an analytical framework that has grown out from the field of border studies with a specific focus on High Asia. This framework is anchored in van Schendel's (2002, 2007) proposal to conceive High Asia under a single rubric, for which he suggests the neologism 'Zomia'.[4] Following this suggestion James Scott (2009; cf. Kreutzmann 2013) has further argued that the various strategies of Zomian communities are often wrongly identified as being primitive (or barbarian) and static (or politically inept). Mobile livelihoods, relatively egalitarian social structures, religious heterodoxy or syncretism,

seen as the political divides that were gradually established around the globe with the onset of modern state-making in the eighteenth century (Baud and Schendel 1997: 213–216). While borders demarcate a state's sovereign territory, the concept of 'borderland' is used to describe regions that are affected by an international border. Baud and van Schendel (1997: 216) have coined the term 'borderland' as an analytical category that encompasses both sides of a state border, since the creation of borders almost always implies "the existence of (new) networks and systems of interaction across them". The third term 'boundary' is particularly used in diplomatic circles when the precise location of an international border – such as between India and China – is being negotiated. The reader needs to be aware that there also is a well-established anthropological understanding of the term 'boundary', namely for denoting the divisions – however permeable – between ethnic and cultural groups (Barth 1969).

[4] Van Schendel derived the neologism 'Zomia' from *zomi*, a term signifying 'highlander' in a number of Chin-Mizo-Kuki languages that are spoken in the borderlands of India, Burma and Bangladesh.

linguistic pluralism, fluid ethnic identities and pliable oral cultures would have allowed them, rather, to effectively negotiate and constantly adjust their positionality vis-à-vis the states or state-like formations with which they engage. In Scott's (2009: 8) words:

> The situation of populations that have deliberately placed themselves at the state's periphery has occasionally been termed, infelicitously (i.e. inappropriately), secondary primitivism. Their subsistence routines, their social organization, their physical dispersal, and many elements of their culture, far from being the archaic traits of a people left behind, are purposefully crafted both to thwart incorporation into nearby states and to minimize the likelihood that state-like concentrations of power will arise among them.

While mountain dwellers in High Asia have often been depicted as barbarians or backward hillbillies, such imaginaries are deceptive. Historically many highland groups have, according to Scott, sought to avoid state burdens such as taxation, conscription, forced labor or adherence to one centrally sanctioned religion without losing the ability to engage in and profit from long-distance trade with lowland societies. And it is precisely for those reasons that mobile livelihoods and fluid ethnic identities are so common amongst them. In other words, highland communities are perhaps better described as "barbarians by design" (Scott 2009: 8), or as a people who craft their trade, identity and mobility as a politically motivated response to state influence. Though scholars have raised questions about the general applicability (Michaud 2010) and empirical accuracy (Formoso 2010; Lieberman 2010; Dove et al. 2011; Karlsson 2013) of Scott's historical analyses, he has nevertheless opened fruitful trajectories in Himalayan research (Shneiderman 2010, 2013; Bergmann 2016). What I advocate here is a perspective that focuses on the role of imperial and national histories in shaping highland communities as well as on the strategies that mountain dwellers have devised to engage with and capitalize on external forces of change. Before addressing these issues in some more detail, it is necessary, however, to say a few more words about the Bhotiyas.

1.2 Who Are the Bhotiyas?

The anthropologist Charles Brown (1992) once published an article with the provocative title 'What we call "Bhotiyas" are in reality not Bhotiyas'. Taken literally, the ethnonym 'Bhotiya' denotes the residents of a region called *Bhot*. The geographic name *Bhot* is etymologically related to the late Sanskrit *Bhotah*, which itself derives from the Tibetan word *Bod*, meaning Tibet (Ramble 1993). According to this rationale, the Bhotiyas are Tibetans. Such an interpretation, however, is questionable with regard to the Bhotiyas in present day Uttarakhand. The name 'Bhotiya' (also 'Bhotia' and 'Bhootia') was popularized during the expansion of the British Empire from Bengal past Nepal to the western Himalaya in the eighteenth and nineteenth centuries (Brown 1987). Its application for a diverse range of Tibetan-speaking Buddhists in Nepal played a crucial role in establishing the association with this religion, and consolidated pejorative connotations of beef eating and alcohol drinking (Ramble 1993). Also in the Kumaon Himalaya British administrators, scholars, and explorers referred to the trans-Himalayan trading groups residing at the northernmost edge of the empire as 'Bhotiyas' (Traill 1992 [1832]; Atkinson 1973 [1882]; Sherring 1974 [1916]). Whilst the adequacy of the term remains disputed amongst both scholars and local residents, I will retain it as a general term of reference, not least because it is the official Scheduled Tribe designation of the people in question.

This Scheduled Tribe of India inhabits several high mountain valleys, all of which are close to international borders in Garhwal and Kumaon, the two administrative divisions and former kingdoms of present-day Uttarakhand. After India won independence from the British Raj in 1947, this region was incorporated into Uttar Pradesh, India's most densely populated state. Only in the year 2000, following decades of political protest, was Uttarakhand carved out as the 27th state of the Indian Union (known as Uttaranchal from 2000 to 2006). In the districts of Uttarkashi and Chamoli of the Garhwal division, reside the so-called Jadh, Tolcha, and Marcha. The two major groups in Kumaon's Pithoragarh District, on whom the present work is focusing, are the Johari and Rang.[5] The Johari reside in the Gori valley, whose upper parts are referred to as Johar and Ralam. The Rang live further east and are split into subdivisions that are also named according to the valleys they inhabit, namely the Darmani (Darma valley), Chaudansi (Chaudans valley) and Byansi (Byans valley).[6] My analysis focuses particularly on the residents of the Darma valley, who have so far received little scholarly attention (see Box 1.1).

The indigenous languages spoken in this area are classified by Grierson (1909) as belonging to the eastern subgroup of the complex pronominalized Himalayan languages of the Tibeto-Burman family. While Grierson applied the term 'Rangkas' only to the dialect spoken in the Johar valley, Sharma (1990; cf. Willis 2007: 21–28) uniformly classifies all the vernaculars spoken by the Kumaoni Bhotiyas with that term. However, most Bhotiyas are multilingual and speak Hindi, Nepali, Pahari and – to a lesser degree and mostly by elderly people – also some Tibetan in addition to their 'mother tongue' (Nawa 2004). In recent years and especially amongst young people also English has gained significance.

Up until the Sino-Indian war began in 1962, the Bhotiyas were actively involved in a trans-Himalayan trade with residents of the Tibetan highland. Sugar, grains, and manufactured-wool products from India were exchanged for salt, wool, animals, and borax from Tibet (Pant 1935).[7] Sheep and goats as well as yaks and yak-crossbreeds were reared and used during the summer months to cross the mountain passes leading towards the Tibetan plateau. In general, low-caste Harijan families (today officially recognized as Scheduled Castes) were, and to some extent still are,

[5] The Bhotiyas of Kumaon, particularly the Johari, also use the ethnonym 'Shauka' when referring to themselves in the Hindi or Nepali language.

[6] Some Byansi also reside in neighboring valleys of far-western Nepal (Nawa 2000).

[7] Borax (also referred to as *tincal*) is a mineral from the saline lakes on the Tibetan plateau. European porcelain manufacturers used it as a mordant in dyeing processes.

Box 1.1 The Darma Valley

The Darma valley is located in Dharchula Thesil (an administrative unit in India that stands below the district level) of the Pithoragarh District in Uttarakhand. The valley has two divisions, namely Malla- and Talla-Darma ('upper' and 'lower' Darma). Malla-Darma comprises the 13 (earlier 14) settlements of the Darmani, which are scattered between altitudes of approximately 2,250 and 3,500 m; Talla-Darma refers to the area below the first summer settlement Sela. Especially the so-called Anwals, who have traditionally worked as shepherds and subordinate traders for the Darmani, inhabit this lower region. Malla-Darma is connected through high passes to the neighboring Johar valley in the west (Ralam pass) and the Byans valley in the east (Shinla pass). The main route from Darma toward the Tibetan plateau was via the Labe Dhura pass. Major trade destinations in Tibet were the marts in Taklakot and Gyanima. Whilst the number of migrating people has steadily declined after the closure of the Indo-Tibetan border in 1962, recent developments have given a new impetus.

The main occupations of migratory households are animal husbandry and crop farming. Reared animals include sheep, goats, cattle (including traditional crossbreeds of yak and cow) and horses. Two varieties of buckwheat, namely *palti* (*Fagopyrum esculentum*) and *bai* (*F. tataricum*) are cultivated extensively in the upper settlements besides potatoes and other vegetables. The extraction and farming of aromatic and medicinal plants marks an increasingly important source of income for the villagers in recent years. The winter settlements of the Darmani are located between Dharchula and Jauljibi along the Kali River: Sipu (Malla Galati, Nayabasti), Marcha (Talla Galati), Tidang (Kalika), Go (Balvakot), Philam (Charchum), Dhakar (Ghatibagad), Bon (Gothi), Son/ Dugtu (Nigalpani), Baling (Galati), Nagling (Galati), Chal (Junibagad, Nayabasti), Sela (Ansingara), Dantu (Jauljibi). In the winter settlements only smaller plots of agricultural land are under cultivation, mainly with rice and wheat. During winter larger flocks of sheep and goats are pastured at the foothills of the Himalaya, mostly around Tanakpur.

closely associated with a Bhotiya household, for whose members they conduct various agricultural tasks and domestic chores.

In the course of the year the Bhotiyas use different ecological belts, each of which exhibits specific potentials and limitations for crop farming, and forest and pasture use (Nüsser 2006; Bergmann et al. 2012: 179). The narrow belt of the Outer Himalaya (*Bhabar*) rises out of the northern parts of the Gangetic Plains (*Terai*), where sub-humid tropical Sal-forests (*Shorea robusta*) are found and utilized as winter pastures up to an altitude of approximately 1,000 m. The Lesser Himalaya (*Himanchal*) is adjoined as a 70–100 km broad belt stretching in a northerly direction and covering an altitude of between 1,500 and 3,000 m. In lower parts this belt is extensively covered by *Pinus roxburghii*-forests, which often exhibit an herbaceous understorey sustained by intentional burnings. In the vicinity of settlements most of the available land is terraced and cultivated for double cropping with rice, wheat, millet and various legumes. The Bhotiyas' winter settlements are located in the northernmost part of this belt. Two of these settlements, Munsiari and Dharchula (called Darchula on the Nepalese side), have developed into densely populated and urbanized centers.

The adjoining zone of the High Himalaya (*Himadri*) is about 30–50 km wide. Glaciated mountain peaks, of which some exceed an altitude of 7,000 m, dominate the scenery. Its narrow transversal valleys give rise to montane forests where evergreen oaks (*Quercus semecarpifolia*, *Qu. floribunda* and *Qu. leucotrichophora*) alternate with areas dominated by conifers (*Abies spectabilis*, *Cupressus torulosa*) or deciduous trees (*Alnus nepalensis*, *Aesculus indica*, *Acer* spp., and *Betula utilis* var. *jacquemontii*). The summer settlements are located in altitudes between approximately 2,400 and 3,500 m. In these settlements different species of buckwheat, barley, potatoes, radish and cabbage are cultivated. Above 3,600 m high altitude grasslands are found, and prevail into the Tibetan Himalaya that forms the northernmost part of Kumaon. Being covered with meadows and dwarf shrubs, the relatively wide and shallow valley bottoms are widely used as summer pastures (*bugyal*). From these parts of the valleys, which are nowadays strictly controlled by the Indian army and Indo-Tibetan Border Police (ITBP), high passes lead toward the Tibetan plateau.

In the following sections I examine the competing forms of power that have unevenly influenced Kumaon's geographies of trade, identity and mobility since colonial times. In doing so I provide insights into how the Bhotiyas have harnessed their endogenous capacities and capitalized on external forces of change whilst negotiating their positionality within this transforming socio-spatial landscape (Bergmann et al. 2008).

1.3 Geographies of Trade, Identity and Mobility in Kumaon

Before the onset of British colonialism in the Himalayan border region, ruling dynasties in the border triangle of present day India, Tibet and Nepal created a regionally arranged power-geometry (Brown 1984; Joshi and Brown 1987). Control over the trans-Himalayan trade was a focus of special attention for these rulers, both in order to regulate movements across the high passes and to absorb some kind of surplus from it. The kingdoms of Garhwal and Kumaon attracted increasing British attention when the Gorkhas, a ruling power from Nepal, conquered them in 1790 (Fig. 1.2). After having defeated the Gorkhas in 1815, the British East India Company established Kumaon as a so-called Non-regulation Province and therewith secured its first direct access route to the Tibetan plateau. At that time the Bhotiyas' trade was limited to the barter of locally needed products, such as grain, salt, cloth and wool (Raper 1812; Moorcroft 1818). Ladakhi traders monopolized the long distance trade in cashmere (or pashmina) wool used for the weaving of fine

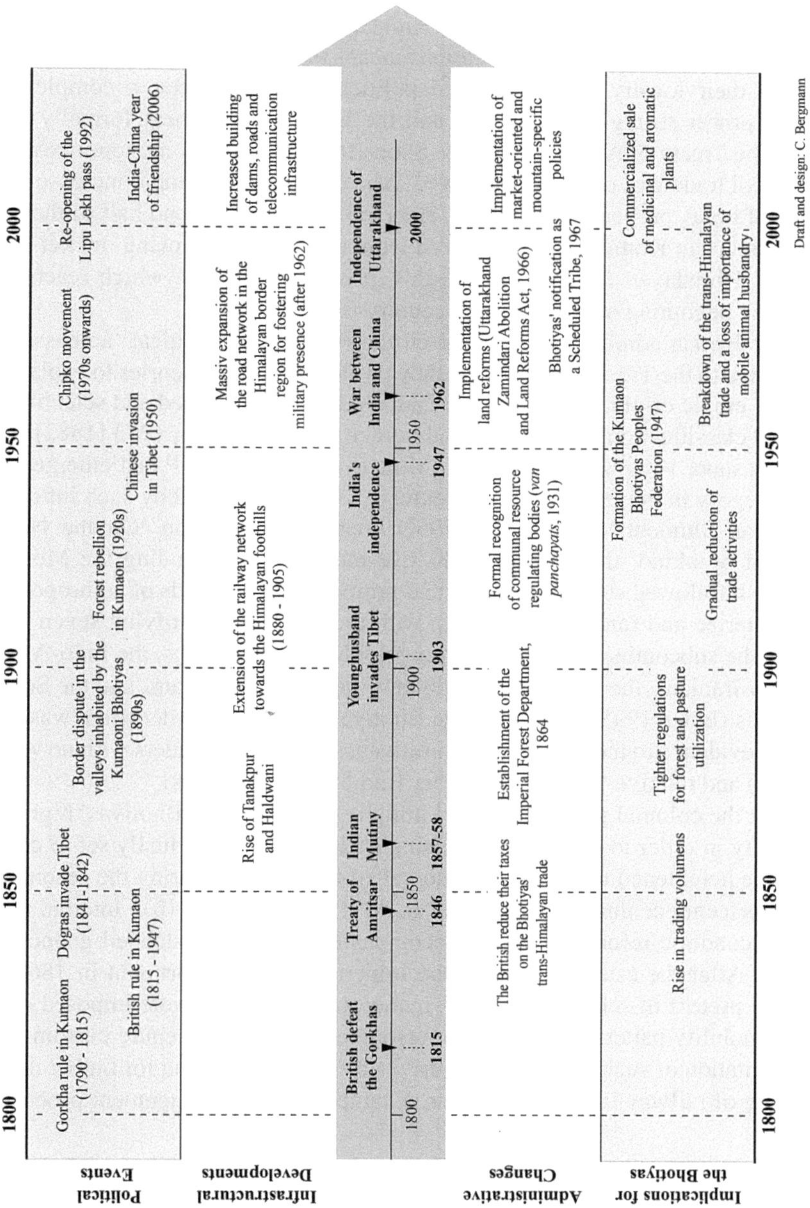

Fig. 1.2 Historical timeline

textiles such as shawls. A primary aim of the British administration in Kumaon was to divert this lucrative trade from Tibet directly into British territory (Rizvi 1999).

The invasion of Ladakh (1834) and, more particularly, western Tibet (1841) by the Dogras – a subsidiary of the powerful Sikh kingdom that rivalled the British in the Punjab – adversely affected the trans-Himalayan trade of the Bhotiyas (Bergmann 2016: 5–8). The British conceived various means to both balance the traders' losses and gain their loyalty in an unsettled political situation. After a complex and dynamic power struggle, the Dogras and the British government formally allied through the Treaty of Amritsar in 1846. Soon after, the Ladakhi monopoly over the shawl-wool trade was officially dissolved. As a consequence a rapid increase in the volume of trade, particularly for wool, emerged during the second half of the nineteenth century in Kumaon. Industrialized wool mills in fast-growing market towns along the Himalayan foothills fuelled this growth in demand, which reached its peak at the beginning of the twentieth century (Roy 2003).

Whilst British administrators relied on the Bhotiyas as political 'ambassadors' and a bridge to the Tibetan authorities, they proposed various theories to explain the Bhotiyas' ethnic origin. Eventually they adopted the standardized and scientifically grounded classification as a 'Mongoloid border tribe' (Atkinson 1973 [1882]: 111–115; Srivastava 1966: 204). The term 'Mongol' or 'Mongoloid' had emerged as a racial category in the late eighteenth century. It was propagated by such influential writings as Blumenbach's (1865 [1795]) doctoral dissertation 'On the Natural Variety of Mankind' that distinguished five major races, including the Mongols. The British followed such approaches and promoted the methods of anthropometry to characterize and rank the colonized, with the result of identifying seven major races on the subcontinent alone: the Mongoloids, the Dravidians, the Indo-Aryans, the Turko-Iranians, the Mongolo-Dravidians, the Aryo-Dravidians, and the Scytho-Dravidians (Bates 1995). In case of the Bhotiyas, the affix 'border tribe' was justified with evidence to their apparent 'primitiveness' (itinerant traders with no written language) and relative 'isolation' (neither Buddhists nor Hindus).

Whilst the colonial government had initially facilitated the Bhotiyas' high level of mobility in order to ensure a flourishing trade, this was gradually set to change amidst the heightened tensions of the so-called Great Game during the second half of the nineteenth century (Bergmann et al. 2011; Bergmann 2016).[8] Internal political and economic reform processes accompanied this tension-loaded geopolitical situation. After the establishment of the Imperial Forest Department in 1864 and under the pretext of scientific forestry, numerous restrictions were imposed on the traders' mobility patterns, strategies of resource use and land tenure customs. The implementation of such measures was driven by a strong demand for timber used in the laying of railway lines in the plains. Examples of the management procedures

[8] During the so-called Great Game, Britain and Russia rivaled for political ascendancy in Central Asia. This conflict evolved after Napoleon's army retreated from Russia in 1813 and lasted approximately until the Russian October Revolution in 1917. The Great Game had several heightened periods, of which the turn of the twentieth century is one. It is at this time that the British invaded Tibet and thereafter established it as a buffer state (Anand 2009).

employed include the assignment of market and halting places (*parao*) as well as the total closure of forested areas to grazing in the Lesser Himalaya. Between 1910 and 1917 also the forested areas around the Bhotiyas' winter settlements were brought under direct control of the Forest Department. In the midst of these broader re-configurations the Bhotiyas' commercial activities began, for a variety of reasons, gradually to dry up from the 1920s onwards (Bergmann et al. 2011: 112).

Following India's independence in 1947, raising political disputes in Tibet further distorted the traders' cross-border mobility, which finally came to a halt after the war between India and China in 1962. With the discontinuation of the trans-Himalayan trade, an important economic cornerstone for the whole region ceased to exist. Whilst the Bhotiyas were deprived from an important means of securing their livelihoods, the Government of India was endeavored to set a positive sign and approved the community's long-standing demand for recognition as a Scheduled Tribe in 1967. Since then the community is entitled to benefit from the various quota-based measures of affirmative action for India's Scheduled Tribes. Resounding economic success in the region was achieved after India's market liberalization in the 1990s and after Uttarakhand was carved out as an independent state in 2000 from Uttar Pradesh, whose government had hitherto ignored demands coming from the Himalayan periphery. The newly established Government of Uttarakhand aimed at combining market-based approaches with a mountain-specific policy, of which hydro-energy, tourism and medicinal plant extraction evolved as the most important building blocks. Amidst these emergent projects and relations within a market-oriented border region the Bhotiyas started to re-fashion their mobility such as to make it once again rewarding in both economic and socio-cultural terms. In other words, the Bhotiyas continued to co-produce and actively negotiate the forces and relationships that transform their socio-spatial landscape.

1.4 Processes of Negotiation in a Transforming Socio-spatial Landscape

The Bhotiyas reside in an area that has not only been overshadowed by competing forms of power, but also by both Buddhist and Hindu cultural influences. The traders' traditional cycles of exchange, and their regular journeys over the high mountain passes, were affected by broader processes of political change and attuned to the calendar of religious festivals, not least because at many fairs religious and commercial activities went hand in hand (Chandola 1987; van Spengen 1998). Moreover, until today rituals are integral to people's seasonal movements and modes of environmental engagement. The different valley communities of the Bhotiyas have made use of their peripheral location and cultural distinctiveness – including their Tibetan-like appearance – to perpetually negotiate their positionality vis-à-vis each other as well as outside agents.

Although numerous Pahari were involved as customers, distributors and suppliers of trade products, they usually had no personal access to the transport routes over the high passes and the Tibetan markets.[9] This exclusion can at least partly be attributed to the performative character of the Bhotiyas' commerce, which was tightly linked to the so-called *mitra*- and *sarji*-systems (Leder 1996).[10] Through these systems enduring trade partnerships were established and the trans-Himalayan flow of commodities and people regulated. Both involved ritualized practices that ignored the important Hindu matter of touch-pollution. The Bhotiyas made use of these practices to assure their lucrative trade monopoly, since they also popularized the half-truth that the Tibetans regarded commensal relations as a necessary requirement for trade (Srivastava 1966). This idea gained plausibility in the light of their physical appearance: from the viewpoint of Kumaon's Hindu population the Bhotiyas presumably descended from yak-eating and alcohol-drinking Tibetans with whom they maintained ritualized personal relations. Other than the difficulties and dangers of traveling and transporting goods over the mountain passes, and the problems resulting from the different languages, the proclaimed character of the trans-Himalayan trade effectively kept away possible Pahari competitors.

At the same time, these ritualized modes of trade were strongly supported by the authorities in Tibet, which might be a reason for the existence of similar practices among other trans-Himalayan trading groups in Nepal (von Fürer-Haimendorf 1975). The Tibetan authorities relied on them to articulate their territorial and fiscal power. On the one hand the *sarjis* were sent as delegates of Tibetan governors. They controlled the high passes in order to prevent the entry of Europeans as well as the spillover of human or animal diseases into Tibet; and they collected dues from the Bhotiyas before they declared ritually the inception of the trade season and opened up the passes. On the other hand the Tibetans utilized the *mitra*-system for directing the different valley communities to particular markets and securing advantageous exchange rates. When toward the end of the nineteenth century the British tried to intervene in these traditional modes of trade in order to establish their own economic machinery, based on the idea of 'free trade' (van Spengen 2000), this resulted in overt tensions. Both the Tibetans and the Bhotiyas were eager to maintain their trade on its 'old footing', i.e. on the basis of customary arrangements. A Tibetan governor even enforced the temporary closure of one of the most frequented high passes (Lipu Lekh) of Kumaon, which effectively frustrated British ambitions (Bergmann 2016: 10).

When commercial activities with the Tibetans began to dry up even before the closure of the border for the reasons outlined above, the Bhotiyas actively renegotiated their livelihood and ethnic identity. In 1947, on the eve of India's independence, the Kumaon Bhotiya People's Federation (KBPF) was founded as a

[9] 'Pahari' literally means 'of or belonging to the mountains' and commonly refers to the majority Hindu hill populations in Nepal and India. These people speak Indo-European languages that are also classified by this term.

[10] The *sarjis* were delegates of Tibetan governors (*jongpen*). Exclusive trade partners were called *mitra*, which literally translates as 'friend'.

political organ through which community members articulated their claims for governmental support and their ideas for a possible diversification of a trade-centered economy. The demand of having better access to forests and grasslands was one of their fundamental aspirations. But calculations regarding their own self-presentation also informed the negotiations, since the Federation's members sensed that access to political and educational resources would be determined by their officially acknowledged ethnicity. Anthropologists have so far discussed these negotiations in terms of Hinduization (Srivastava 1966), i.e. an acculturative process involving the Bhotiyas' assertion of a high-caste Hindu identity, despite the pervasive perception that they are actually 'crypto-Buddhists' (Bergmann et al. 2008).

However, the term 'crypto-Buddhist' itself implies not only a barrier to Hinduization, but also a practical strategy employed earlier by the Bhotiyas to secure a lucrative trade monopoly, and later to gain special governmental treatment, for which the identification and public display of distinguishing socio-cultural resources such as certain rituals and customary institutions was crucial. Right from the beginning the members of the KBPF skillfully assembled widespread representations that public authorities had coined of their community. An example would be their classification as a Mongoloid border tribe, including associated features such as egalitarian social structures, gender equality and sexual liberty. In using and converting such representations for their own purposes, these actors articulated their command over emergent scales of social activity, particularly the national, whilst trying to affirm the Bhotiyas' entitlement to Scheduled Tribe status. Although this claim gained political attention only after the Sino-Indian war in 1962, the Bhotiya's official recognition as a Scheduled Tribe of India marks the federation's greatest success.

ST-status provided leeway to community members searching for alternative livelihoods, such as shop-keeping or government employment (Prasad 1989; Hoon 1996). However, seasonal mobility did not lose its appeal altogether. Whereas the number of migratory households decreased rapidly since the 1960s, this trend should not be considered as irreversible. This is indicated by the fact that seasonal migration has in recent years experienced a new impetus. In the course of ongoing state and market reforms income opportunities such as tourism have opened up in the valleys of the Bhotiyas. Young and educated community members particularly welcome these opportunities. At the same time mobility related rituals have been re-discovered and institutionalized as regular gatherings to which all members of a village may regularly visit their summer settlement (Bergmann et al. 2011).

1.5 Methodological Approach

This book draws from recurring phases of fieldwork in North India. These were conducted between 2008 and 2012 and extended over a total period of approximately 24 months. Field research was mainly carried out in the Darma valley and focused on the historical developments in trans-Himalayan trade, identity politics and the mobile practices of the Bhotiyas. I therefore relied heavily on intense and systematic participant observation. I took an active part in seasonal migration, accompanied shepherds to the alpine pastures and on their extended journeys to and fro between the Darma valley and the Outer Himalaya, accomplished daily work activities with villagers in the fields and forests, and documented numerous rituals and ceremonial events.

Through loosely structured and repeated interviews, I collected life stories and oral histories, especially from ritual specialists and village elders. Further interviews with a range of people of both sexes and of different castes in most of the upper settlements then allowed me to gain a broader perspective on specific issues, ranging from the negotiations over Scheduled Tribe status, informal institutions, ritual practices and the changing modalities of forest and pasture use. For comparing and contrasting the qualitative information base thus obtained with the circumstances of people in the adjoining valleys I realized shorter research stays in several villages of the Chaudans and the lower parts of the Byans valley. In Dharchula, Munsiari, Pithoragarh, Nainital, Haldwani Lucknow and Delhi I realized various informal discussions with Bhotiyas of different valleys, out-migrated persons, government officials and members of other regional groups. In order to "reflexively approach the past and reconsider the present" (Ladwig et al. 2012), especially in terms of the region's territorial restructuring, I further undertook archival fieldwork for a total period of 3 months, namely in the Kumaon Regional Archive in Nainital and the Uttar Pradesh State Archive in Lucknow.

If permitted, one-on-one interviews as well as group discussions and oral recitations during rituals were captured with a digital voice recorder. If the situation was not suitable for such a recording I took notes either during or shortly after a conversation had taken place. I also made extensive use of photography and video recording to visualize and explore certain forms of knowledge and practice. The recorded visual material often served as an entry-point for further discussion. For the most part I conducted research in the Hindi language, which nearly all residents of the region speak fluently. In other instances my local field assistant provided me with on-the-fly translations as a conversation was unfolding. Ritual performances were almost always recorded in the Darma language and afterwards transcribed and translated into Hindi.

1.6 How the Book Is Organized

The rationale for the overall structure of the present book arises from the twin objective to comprehend the centrality of the Himalayan border region within the entangled imperial and national histories of adjacent states as well as to explore the agency of its marginalized residents in dealing proactively with shifting forms of territorial boundedness. The book builds upon the premise that larger processes of territorial and economic restructuring are not only determined by policy decisions of elites in centers of supremacy, but are also shaped by the countervailing strategies of marginalized communities (Ludden 2011). Through an engagement with current debates on borderlands and transboundary environments I examine how the Bhotiyas have harnessed their endogenous capacities and capitalized on external forces of change whilst negotiating their positionality within Kumaon's transforming socio-spatial landscape. The empirical material highlights the multifaceted repertoires of practice and knowledge that constitute their "borderland advantage" (Flynn 1997: 312). This theme is developed in three main chapters.

Chapter 2 explores how the Bhotiyas have dealt with imperial forms of power and authority in order to realize their trans-Himalayan trade under British colonial rule in nineteenth-century Kumaon. Based on a combined analysis of oral and written historical sources, I show that British imperial sovereignty, similar to that of older regimes in the area, remained malleable and contested within the wider relational field of the Bhotiyas' trans-Himalayan trade. When the British East India Company annexed Kumaon in 1815 it was recognized as a so-called Non-regulation Province, which meant that government officials could flexibly interpret executive orders to suit the realities on the ground. Procedural simplicity and discretionary decisions created scope for both shaping and contesting British hegemony, leading to an adaptive transformation of imperial rule. Through a close examination of British interactions with Kumaon's traders, the chapter reveals the frictions that arose from this exceptional legal status. This focus serves to address the broader question of how sovereign claims work through multiple and shifting articulations, from frontier narratives to cartographic representations and from fluid relationships of allegiance to fixed state boundaries. The analysis considers a previous call to conceive High Asia as a continuous zone and an agentive site of political action by arguing that confluent territories and overlapping sovereignties are key to understanding the development of trans-Himalayan trade in this imperial environment.

Starting on the eve of India's independence in 1947, the objective of Chap. 3 is to turn attention to the scalar dimensions and repercussions of the Bhotiyas' struggle for tribal status. Whilst examining the multiple and contradictory meanings of scale, scholars have extended their arguments beyond its conventional conception as fixed and hierarchically ordered spatial containers. In a growing body of literature scales are considered within strategies that actors deploy in pursuing their various agendas. For example, whilst some actors favor localization to effectively realize their interests others might opt to regionalizing, nationalizing, or globalizing their strategies. However, the question as to what extent a contentious scalar politics is implicit within processes of ethnic identification has until now largely escaped consideration. In addressing this shortcoming of contemporary research, I analyze how the Bhotiyas have leveraged the performativity of scale through various narratives and social practices in order to assert who they are or want to be in the context of two macro-institutional transformations: the shift from British imperial rule to India's nation-state sovereignty and the creation of a new mountain state, viz. Uttarakhand, within that country's federal system of government. As such, this research will have broader relevance to other multi-ethnic border regions in and beyond High Asia.

A detailed analysis of the Bhotiyas' mobile practices in the market-oriented border region of today's Uttarakhand is then provided in Chap. 4. A wide range of ethnographic material is presented to consider how the Bhotiyas did re-fashion their pastoral mobility, including associated rituals, following India's implementation of liberal economic reforms since the early 1990s. Drawing from Tim Ingold's (2000) so-called 'dwelling perspective', which foregrounds how actors become involved with and get to know their environs practically, I argue that rituals are integral to people's overall work activities. They allow them to incorporate the 'seen' and 'unseen' constituents of the environment into rhythmic patterns of movement that become embedded in social relations and embodied in a variety of skills. Following this argument it is shown how rituals foster people's responsiveness to the perceived fluctuations of their dwelt-in environments: the seasonal cycle of weather conditions, the vegetative growth cycles of plants, and the life cycles of fellow human and other animal creatures. However, while ritual and myth convey this responsiveness in kinaesthetic and poetic ways, the Bhotiyas' pastoral mobility is always already mixed with politics. In this politics, care occurs alongside exploitation, judgment alongside ignorance, skilled practice alongside routine operation, and resonance alongside resistance. The book closes with an epilogue, which summarizes the main arguments of the study and identifies possible directions for follow-up research.

References

Alvarez RR (1995) The Mexican-US border: the making of an anthropology of borderlands. Annu Rev Anthropol 24:447–470

Anand D (2009) Strategic hypocrisy: the British imperial scripting of Tibet's geopolitical identity. J Asian Stud 68(1):227–252

Anderson J, O'Dowd L (1999) Borders, border regions and territoriality: contradictory meanings, changing significance. Reg Stud 33(7):593–604

Atkinson ET (1973 [1882]) The Himalayan gazetteer, vol III, part I. Natraj Publishers, Dehra Dun [originally published under the title 'The Himalayan districts of the north western provinces of India']

Baghel R, Nüsser M (2015) Securing the heights: the vertical dimension of the Siachen conflict between India and Pakistan in the eastern Karakoram. Political Geogr 48:24–36

Barth F (1969) Introduction. In: Barth F (ed) Ethnic groups and boundaries: the social organisation of culture difference. George Allen and Unwin, London, pp 9–38

Bates C (1995) Race, caste and tribe in central India. The early origins of Indian anthropometry. In: Robb P (ed) The concept of race in South Asia. Oxford University Press, New Delhi, pp 219–259

Baud M, van Schendel W (1997) Toward a comparative history of borderlands. J World Hist 8(2):211–242

Bergmann C (2016) Confluent territories and overlapping sovereignties: Britain's nineteenth-century Indian empire in the Kumaon Himalaya. J Hist Geogr 51:88–98

Bergmann C, Gerwin M, Nüsser M, Sax WS (2008) Living in a high mountain border region: the case of the 'Bhotiyas' of the Indo-Chinese border region. J Mt Sci 5:209–217

Bergmann C, Gerwin M, Sax WS, Nüsser M (2011) Politics of scale in a high mountain border region: being mobile among the Bhotiyas of the Kumaon Himalaya, India. Nomadic Peoples 15(2):104–129

Bergmann C, Gerwin M, Nüsser M, Sax WS (2012) State policy and local performance: pasture use and pastoral practices in the Kumaon Himalaya. In: Kreutzmann H (ed) Pastoral practices in High Asia: agency of 'development' effected by modernization, resettlement and transformation. Springer, Heidelberg, pp 175–194

Berreman GD (1963) Hindus of the Himalayas: ethnography and change. University of California Press, Berkeley

Blumenbach JF (1865 [1795]) On the natural variety of mankind. In: Bendyshe T (ed and transl) The anthropological treatises of Johann Friederich Blumenbach with memoirs of him by Mary and Flourens, and an account of his anthropological museum by R. Wagner, and the inaugural dissertation of John Hunter on the varieties of man, 3rd edn. Longman, Green, Longman, Roberts & Green, London, pp 146–276

Brown CW (1984) The goat is mine, the load is yours: morphogenesis of 'Bhotiya-Shauka', U.P., India. Lund University, Department of Social Anthropology, Lund

Brown CW (1987) Ecology, trade and former Bhotiya identity. In: Raha MK (ed) The Himalayan heritage. Gian Publishing House, New Delhi, pp 125–138

Brown CW (1992) What we call 'Bhotiyas' are in reality not Bhotiyas: perspectives of British colonial conceptions. In: Joshi MP, Fanger AC, Brown CW (eds) Himalaya: past and present, vol II. Shree Almora Book Depot, Almora, pp 147–172

Chandola K (1987) Across the Himalayas through the ages: a study of relations between central Himalayas and western Tibet. Patriot Publishers, New Delhi

Dove MR, Jonsson H, Aung-Thwin M (2011) Debate. The art of not being governed: an anarchist history of upland Southeast Asia by James C. Scott (Review). Bijdragen tot de Taal-, Land- en Volkenkunde 167(1):86–99

Ehlers E, Kreutzmann H (2000) High mountain ecology and economy: potential and constraints. In: Ehlers E, Kreutzmann H (eds) High mountain pastoralism in northern Pakistan. Steiner, Stuttgart, pp 9–36

Flynn DK (1997) 'We are the border': identity, exchange, and the state along the Benin-Nigeria border. Am Ethnol 24(2):311–330

Formoso B (2010) Zomian or zombies? What future exists for the peoples of the Southeast Asian Massif? J Glob Hist 5(2):313–332

Gellner D (2013) Northern South Asia's diverse borders, from Kachchh to Mizoram. In: Gellner D (ed) Borderland lives in northern South Asia (with an afterword by Willem van Schendel). Duke University Press, Durham, pp 1–23

Gerwin M, Bergmann C (2012) Geopolitical relations and regional restructuring: the case of the Kumaon Himalaya, India. Erdkunde 66(2):1–17

Grierson GA (1909) Linguistic survey of India, vol III. Tibeto-Burman Family. Superintendent of Government Press, Calcutta

Harris T (2013) Geographical diversions: Tibetan trade, global transactions. University of Georgia Press, Athens

Hoon V (1996) Living on the move: Bhotiyas of the Kumaon Himalaya. Sage, New Delhi

Ingold T (2000) The perception of the environment. Essays in livelihood, dwelling and skill. Routledge, London

Johnson C, Jones R, Paasi A, Amoore L, Mountz A, Salter M, Rumford C (2011) Interventions on rethinking 'the border' in border studies. Political Geogr 30(2):61–69

Jones R (2012) Spaces of refusal: rethinking sovereign state power and resistance at the border. Ann Assoc Am Geogr 102(3):1–15

Joshi MP, Brown CW (1987) Some dynamics of Indo-Tibetan trade through Uttarakhand (Kumaon-Garhwal), India. J Econ Soc Hist Orient 30(3):303–317

Karlsson BG (2013) Evading the state: ethnicity in northeast India through the lense of James Scott. Asian Ethnol 72(2):321–331

Kohli A (2006) Politics of economic growth in India, 1980–2005, part II. The 1990s and beyond. Econ Polit Wkly 41(14):1361–1370

Kreutzmann H (2013) Boundary-making as a strategy for risk reduction in conflict-prone spaces. In: Müller-Mahn D (ed) The spatial dimension of risk: how geography shapes the emergence of riskscapes. Routledge, London, pp 154–171
Ladwig P, Roque R, Tappe O, Kohl C, Bastos C (2012) Fieldwork between folders: fragments, traces, and the ruins of colonial archives. Max Planck Inst Soc Anthrop Work Pap 141:1–27
Leder S (1996) Salz für Getreide: Aus dem Leben Tibeto-Indischer Grenzhändler. In: Emmer G, Mückler H (eds) Alltagskulturen in Indien: Aktuelle Entwicklungen in der indischen Gesellschaft. IKO-Verlag für Interkulturelle Kommunikation, Frankfurt am Main, pp 79–96
Lieberman V (2010) A zone of refuge in Southeast Asia? Reconceptualizing interior spaces. J Glob Hist 5(2):333–346
Ludden D (2011) The process of empire: frontiers and borderlands. In: Bang PF, Bayly CA (eds) Tributary empires in global history. Palgrave Macmillan, London, pp 132–150
Malone DM (2011) Does the elephant dance? Contemporary Indian foreign policy. Oxford University Press, Oxford
Michaud J (2010) Editorial – zomia and beyond. J Glob Hist 5(2):187–214
Michaud J (2011) Conclusion: lesson for the future. In: Michaud J, Forsyth T (eds) Moving mountains: ethnicity and livelihoods in highland China, Vietnam, and Laos. UBC Press, Vancouver, pp 215–227
Moorcroft W (1818) A journey to lake Manasarovar in Undes, a province of little Tibet. Asiat Res 12:380–536
Nawa K (2000) Ethnic categories and their usages in Byans, far western Nepal. Eur Bull Himal Res 18:36–57
Nawa K (2004) Language situation and 'mother tongue' in Byans, far western Nepal. Stud Nepali Hist Soc 9(2):261–291
Nüsser M (2006) Ressourcennutzung und nachhaltige Entwicklung im Kumaon-Himalaya (Indien). Geogr Rundsch 58(10):14–22
Paasi A (2009) Bounded spaces in a 'borderless world': border studies, power and the anatomy of territory. J Power 2(2):213–234
Pant SD (1935) The social economy of the Himalayas. Based on a survey in the Kumaon Himalayas. George Allen and Unwin, London
Prasad RR (1989) Bhotia tribals of India. Dynamics of economic transformation. Gian Publishing House, New Delhi
Ramble C (1993) The name Bhotey. Himalayan 6(5):17
Raper FV (1812) Narrative of a survey for the purpose of discovering the source of the Ganges. Asiat Res 11:446–564
Rizvi J (1999) The trade in pashm and its impact on Ladakh's history. In: van Beek M, Bertelsen KB, Pedersen P (eds) Ladakh: culture, history and development between Himalaya and Karakoram. Aarhus University Press, Aarhus, pp 317–338
Roy T (2003) Changes in wool production and usage in colonial India. Mod Asian Stud 37(2):257–286
Rumford C (2006) Introduction. Theorizing borders. Eur J Soc Theory 9(2):155–169
Sanwal RD (1976) Social stratification in rural Kumaon. Oxford University Press, New Delhi
Sax WS (2002) Dancing the self: personhood and performance in the Pandav Lila of Garhwal. Oxford University Press, New York
Sax WS (2009) God of justice: ritual healing and social justice in the central Himalayas. Oxford University Press, New York
Scott JC (2009) The art of not being governed. An anarchist history of upland Southeast Asia. Yale University Press, New Haven
Sharma DD (1990) Tibeto-Himalayan languages of Uttarakhand. Mittal Publications, New Delhi
Sherring CA (1974 [1916]) Western Tibet and the British borderland. The sacred country of Hindus and Buddhists. With an account of the government, religion and customs of its peoples. Cosmo Publications, Delhi

Shneiderman S (2010) Are the central Himalayas in zomia? Some scholarly and political considerations across space and time. J Glob Hist 5(2):289–312
Shneiderman S (2013) Himalayan border citizens: sovereignty and mobility in the Nepal-Tibetan Autonomous Region (TAR) of China border zone. Political Geogr 35:25–36
Shneiderman S (2015) Rituals of ethnicity: Thangmi identities between Nepal and India. University of Pennsylvania Press, Philadelphia
Srivastava RP (1966) Tribe-caste mobility in India and the case of Kumaon Bhotias. In: von Fürer-Haimendorf C (ed) Caste and kin in Nepal, India and Ceylon. Asia Publishing House, Bombay, pp 161–212
Sturgeon JC (2004) Border practices, boundaries, and the control of resource access: a case study from China, Thailand and Burma. Dev Chang 35(3):463–484
Traill GW (1992 [1832]) Statistical report on the Bhotiya mehals of Kumaon. In: Joshi MP, Fanger AC, Brown CW (eds) Himalaya: past and present, vol II. Shree Almora Book Depot, Almora, pp 99–154
Turner FJ (1894) The significance of the frontier in American history. Adam Matthew Digital, Marlborough, pp 197–227 [online resource]
van Schendel W (2002) Geographies of knowing, geographies of ignorance: jumping scale in Southeast Asia. Environ Plan D Soc Space 20(6):647–668
van Schendel W (2005) The Bengal borderland: beyond state and nation in South Asia. Anthem Press, London
van Schendel W (2007) The Wagah syndrom: territorial roots of contemporary violence in South Asia. In: Basu A, Roy S (eds) Violence and democracy in India. Seagull Books, Oxford, pp 36–82
van Spengen W (1998) On the geographical and material contextuality of Tibetan pilgrimage. In: McKay A (ed) Pilgrimage in Tibet. Curzon Press, Richmond, pp 35–51
van Spengen W (2000) Tibetan border worlds. A geohistorical analysis of trade and traders. Kegan Paul International, London
Vandergeest P, Peluso NL (1995) Territorialization and state power in Thailand. Theory Soc 24(3):385–426
von Fürer-Haimendorf C (1975) Himalayan traders: life in highland Nepal. Cox and Wyman Ltd, London
Walker A (1999) The legend of the golden boat: regulation, trade and traders in the borderlands of Laos, Thailand and Burma. Curzon Press, Richmond
Warikoo K (ed) (2009) Himalayan frontiers of India: historical, geo-political, and strategic perspectives. Routledge, London
Wendl T, Rösler M (1999) Frontiers and borderlands: the rise and relevance of an anthropological research genre. In: Rösler M, Wendl T (eds) Frontiers and borderlands: anthropological perspectives. Peter Lang, Frankfurt am Main, pp 1–27
Willis CM (2007) A descriptive grammar of Darma: an endangered Tibeto-Burman language. University of Texas, Austin, PhD-dissertation
Wilson TM, Donnan H (2012) Borders and border studies. In: Wilson TM, Donnan H (eds) A companion to border studies. Wiley-Blackwell, Chichester, pp 1–25
Zutshi C (2010) Rethinking Kashmir's history from a borderlands perspective. Hist Compass 8(7):594–608

Chapter 2
Trans-Himalayan Trade in an Imperial Environment

Abstract This chapter explores how the Bhotiyas dealt with imperial forms of power and authority in order to realize their trade under British colonial rule in nineteenth-century Kumaon. Based on a combined analysis of oral and written historical sources I show that British imperial sovereignty, similar to that of older regimes in the area, remained malleable and contested within the wider relational field of the Bhotiyas' trans-Himalayan trade. When the British East India Company annexed Kumaon in 1815 it was recognized as a so-called Non-regulation Province, which meant that government officials could flexibly interpret executive orders to suit the realities on the ground. Procedural simplicity and discretionary decisions created scope for both shaping and contesting British hegemony, leading to an adaptive transformation of imperial rule. Through a close examination of British interactions with Kumaon's traders, the chapter will reveal the frictions that arose from this exceptional legal status. This focus serves to address the broader question of how sovereign claims work through multiple and shifting articulations, from frontier narratives to cartographic representations and from fluid relationships of allegiance to fixed state boundaries. The analysis considers a previous call to conceive High Asia as a continuous zone and an agentive site of political action by arguing that confluent territories and overlapping sovereignties are key to understanding trans-Himalayan trade in an imperial environment.

Keywords British Empire • Trans-Himalayan trade • Bhotiyas • Territoriality • Sovereignty • Non-regulation province

During its first great surge of territorial expansion between 1770 and 1830, the British East India Company made several attempts to access the Tibetan plateau for strategic, economic as well as scientific purposes (MacGregor 1970).[1] Warren Hastings, the first governor general of India, set the ball rolling in 1774 when delegating his private secretary, George Bogle, to lead an initial expedition from Bengal to Shigatse (Markham 1876). In 1783 Hastings built on the first mission's success by appointing his cousin, lieutenant Samuel Turner, to further develop relations

[1] Major parts of this chapter have previously been published in Bergmann (2016).

C. Bergmann, *The Himalayan Border Region*, Advances in Asian Human-Environmental Research, DOI 10.1007/978-3-319-29707-1_2

with the political elites of this remote mountain polity (Turner 1800).[2] However, numerous obstacles, including active opposition from Tibetan aristocrats, frustrated any lasting access. The central Himalayan kingdoms of Garhwal and Kumaon attracted increasing British attention as gateways to Tibet after the Gorkhas, a ruling power from Nepal, conquered them in 1790.

In 1814 the British East India Company waged war on the Gorkhas, whose defeat was sealed with the Treaty of Segauli the following year (ratified in 1816) (Pemble 2008: 359–361). The western half of the Garhwal kingdom was returned to the former ruling dynasty that then staffed the newly established princely state of Tehri Garhwal. The Company, however, annexed the significantly larger part to the east, which was renamed as British Garhwal and Kumaon (hereafter referred to as Kumaon), and recognized as a so-called Non-regulation Province. This legal status was maintained after the region's incorporation into the larger administrative unit of the North-western Provinces in 1836, whereby Kumaon became subsumed under the overall jurisdiction of a lieutenant governor. During the second half of the nineteenth century the region gradually transitioned towards a more formal rule of law, a process that was concluded only in 1925 (Mittal 1986: 25).

This chapter explores how the Bhotiyas negotiated and co-produced the forces through which the Himalayan border region of Kumaon evolved and shifted its shape as an imperial environment throughout the nineteenth century (Fig. 2.1). Based on a combined analysis of oral and written historical sources I show that British imperial sovereignty, similar to that of older regimes in the area, remained malleable and contested within the wider relational field of the Bhotiyas' trans-Himalayan trade.[3] When the British East India Company moved into these remote and inaccessible (tribal) areas, first-hand experiences of local conditions confounded earlier images of India as a uniform agrarian society (Guha 1999: 130–149). In some places, in order to effectively deal with changing local contexts the colonial government suspended the full application of imperial law, complex bureaucratic procedures and separation of powers. Non-regulation Provinces such as Kumaon were administered not by the rule of law but rather "by discretion or executive interposition" (Misra 1977: 81). In practical terms this meant that the commissioners of such provinces were much more powerful than their counterparts in the regulated areas. Whilst being expected to orient their rule toward the applicable acts and regulations of the colony, these government officials flexibly interpreted executive orders to suit the realities on the ground. Procedural simplicity and discretionary decisions created scope for both shaping and contesting British hegemony, leading to an adaptive transformation of imperial rule.

Through a close examination of British interactions with Kumaon's trans-Himalayan traders, I will reveal the frictions that arose from the exceptional legal status of the area as a Non-regulation Province. This serves to address the broader question of how sovereign claims work through everyday and multiple articulations,

[2]For scholarly accounts on these early explorations of Tibet see Bishop (1989: 25–64) and Teltscher (2006).

[3]For a synopsis of recent scholarly debates on the issue of sovereignty see McConnell (2013).

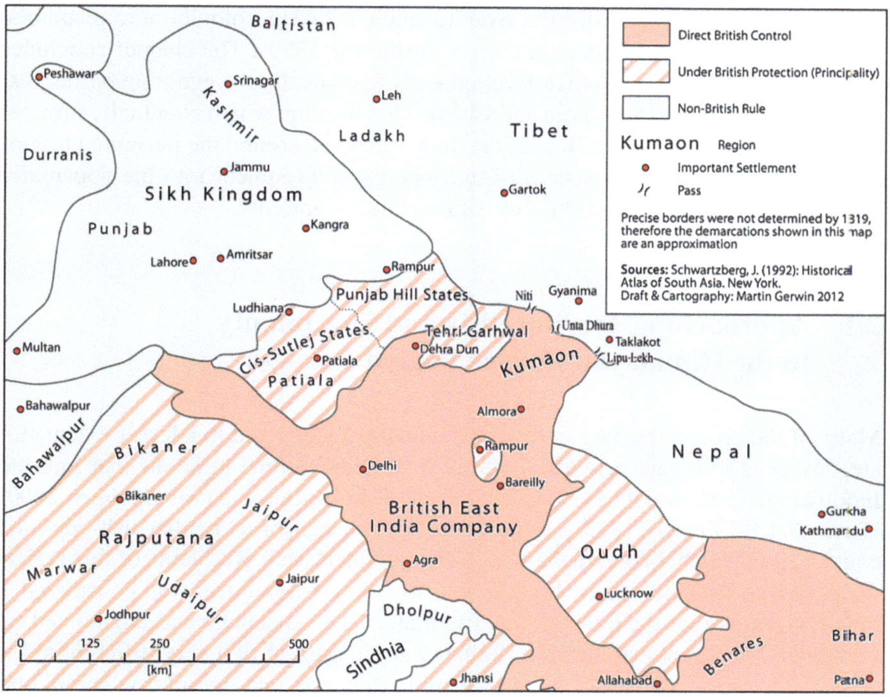

Fig. 2.1 Kumaon's location within the power constellations of 1819 (Bergmann 2016: 93)

from frontier narratives to cartographic representations and from fluid relationships of allegiance to fixed state boundaries. The empirical analysis proceeds in five parts. Departing from an oral narrative about the onset of trans-Himalayan trade in the Darma valley, the first part analyzes the graded and overlapping shades of political power and authority within which the Bhotiyas were positioned before encountering European colonization. How their trade eventually engaged British attention is discussed in the second part, focusing particularly on the political ramifications of an unauthorized journey via the Kumaon and Garhwal Himalaya to western Tibet that William Moorcroft, a veterinary surgeon of the East India Company, conducted in 1812. Following the East India Company's military victory over the Gorkhas in 1815, the third part scrutinizes the strategic efforts of colonial administrators to derive economic and political benefits from the new territorial gain in Kumaon. In the fourth part attention is then turned to the missions of the so-called Pundits, Bhotiyas who secretly surveyed Tibet for the Great Trigonometrical Survey.

The resulting maps played a key role as to how British officials imagined their political authority, namely as being defined territorially. However, toward the end of the nineteenth century it gradually transpired that there still existed a great uncertainty as to who was in control over what and where. In connection with this uncertainty, the fifth part considers the negotiations over territoriality and sovereignty that were triggered by a dispute over taxation between colonial administrators, Tibetan authorities and Kumaoni traders during the 1890s. The chapter concludes with an outlook on postcolonial developments, specifically the evolving border dispute between independent India and China. This dispute, which eventually brought a halt to the Bhotiyas' trans-Himalayan trade, revolved around the intriguing task of translating the anomalous spaces of an imperial environment into the apparently much more clear-cut geographies of nation-state sovereignty.

2.1 Approaching Territoriality and Sovereignty in the Himalayan Border Region

Many of the groups residing in the High Himalaya were, and to a lesser extent still are, involved in a vast network of cultural and commercial relations that link the Indo-Gangetic plain with the Tibetan plateau and Central Asia (van Spengen 2000; Harris 2013).[4] Fisher (1978: 1) introduced the concept of a "neither-fish-nor-fowl contact zone" to characterize the 'Indo-Tibetan interface'.[5] Residents of this region were viewed as Janus-faced brokers whose cultural flexibility was considered a major advantage for the trans-Himalayan trade.[6] This advantage was augmented by the traders' geographical location. Situated at the crux of two very different agro-ecological zones – the arid Tibetan plateau in the north and the subtropical middle hills of the Lesser Himalaya in the south – they played a central role in facilitating commerce for both sides. In other words, it was accepted that altitude, ecology and culture together shaped the complex trading patterns upon which groups such as the Bhotiyas relied.

While scholars from various disciplines have embraced the prospect of such a laboratory-like setting, the model has one major shortcoming: it depicts trans-Himalayan traders as if their lives and livelihoods developed in relative isolation from broader processes of political change. To avoid this shortcoming a gradual rethinking has taken place in recent years. Much of the discussion is anchored in van Schendel's (2002) proposal to conceive High Asia under a single rubric, for which he suggests the neologism 'Zomia'. Although this 'world area' covers large

[4] Initial studies on the resulting agro-trader-pastoralist economies were primarily conducted in northeast and central Nepal. See, for instance, von Fürer-Haimendorf (1964) on the Sherpa, Pignède (1993 [1966]) on the Gurung, Hitchcock (1966) on the Magar and Manzardo (1978) on the Thakali.

[5] For a critical discussion of the concept 'Indo-Tibetan interface' see Shneiderman (2010: 296–300).

[6] The Janus-faced character was most prominently discussed with reference to Goffman's (1959) idea of 'impression management'. Fisher (1986: 96), for instance, argues that the people of Tarangpur in Nepal's Dolpo district are "brokers of goods […], blockers of ideas" in that they adjust themselves to different cultural frames whilst hiding what might be called their core identity. More recently, Shneiderman (2015) has enlarged such a perspective in order to analyze how people perform and shape their identity in relation to multiple nation-state frames.

parts of Central (Inner), South, East and Southeast Asia, it had become a blind spot in their histories. To counter this scholars are now trying to envision the border regions of High Asia as a continuous zone "rather than as disconnected spaces at the peripheries of individual nation-states" (Shneiderman 2013: 28). Scott (2009: 43–44), for instance, analyzes how the realization of mountain livelihoods in Southeast Asia was historically tied not only to dynamic processes of state formation in the lowland centers, but also geared toward the creation and maintenance of ungoverned non-state spaces in the uplands. Scott's version of Zomia dismantles the dominant paradigm of Southeast Asian scholarship as a state-centric vision which neglects the manifold histories, forms of agency and political agendas of those residing in the mountains.

Scott understands upland Southeast Asia historically as a 'non-state space'. People who tried to evade state rule took advantage of altitude, poor infrastructure and other 'frictions of terrain', which they amplified by adopting mobile livelihoods and fluid identities (Scott 2009: 43–44). Even the widespread imperialist mode of indirect rule – whereby authority is delegated from the colonial power to a native ruler – becomes difficult in settings where people are always equipped to engage in further dispersal and scattering.[7] Whilst the Zomia framework has given new impetus to Himalayan scholarship, Shneiderman argues that South Asia's high mountain border zone should not be seen as a 'non-state space'. Instead, she speaks of "a 'multiple-state space', comprised of the territory of all of the nations and states in question, yet transcending the individual sovereignty of any single state" (Shneiderman 2013: 28). In order to advance this perspective I argue that confluent territories and overlapping sovereignties are key to understanding imperial frontiers in the Himalayan region.

Overlapping sovereignties occur when different actors claim control over the same sort of thing, for instance trade. Whilst these claims are often imagined as being inherently territorial, their articulation can also, and at the same time, be marked by a great uncertainty as to who controls where. Without the presence of definite boundaries, actors can effectively negotiate their political authority within such confluent territories through power relationships of allegiance. Although the onset of European colonialism added a new impetus to the negotiations over sovereignty and territoriality, others continued to stake claims: imperial powers such as China or Russia as well as independent polities such as Nepal or Tibet. 'Zomia-thinking' thus foregrounds the agentive participation of multiple states in producing and reproducing sovereign power (Shneiderman 2013: 28). Little consideration, however, has been given to the plural array of processes and actor constellations that contribute to the sovereign articulations of even one state.

Based on the assumption that there is no route to a pure and invariably territorial kind of de jure or legal sovereignty, Agnew (2005: 438) approaches this plural character of sovereignty by distinguishing four sovereignty regimes. These regimes, which account for a state's capacity "to exercise de facto sovereignty", are classified as classic (an effective central state political authority within a well-defined territory);

---

[7] For the modalities of indirect rule see Metcalf (2007) and Mantena (2010).

imperialist (a weak state authority that is characterized by a high degree of external dependency and a constant threat to its territoriality); integrative (different tiers of government are established within complex territorial arrangements); and globalist (a strong central state authority that exercises effective sovereignty beyond its national state territory and enrolls other states into its strategies). For Agnew these regimes represent ideal types. In order to meet their political agendas, states can employ versions of them simultaneously by relying, for instance, on some form of classic sovereignty to manage internal affairs and on globalist sovereignty when dealing with other states and international actors. Such a pluralist understanding of sovereignty is echoed in a number of influential studies on British India's imperial history.

As Legg (2009, 2014) shows in his stimulating examination of the entanglements between imperialism and internationalism during the interwar period, sovereignty in British India was not only highly adaptive but also shaped by tensions across various scales. After India joined the League of Nations as the only non-self-governing member in 1919, its new international status exposed several anomalies with regard to both its external, or imperial, and internal, or colonial, regime of sovereignty. Such anomalies also feature prominently in Benton's (2010) analysis of empires' legal geographies in terms of 'corridors' and 'enclaves'. Imperial administrators often had disputes amongst themselves about the nature and limits of delegated authority. Specific environments such as oceans and mountains triggered numerous legal challenges and created jurisdictional conflicts that complicated imperial rule. Resembling Scott's analysis, Benton also considers lowlanders' depictions of mountains as uncivilized and unruly regions. But she is less interested in the political agency of the colonized mountain dwellers than in the deliberations of imperial jurists who had to specify and justify the exceptional legal status that these regions often received. According to Benton (2008), the perceived low level of development in mountains convinced British officials to incorporate them only partially as 'quasi-sovereign' polities. Benton's prime example for quasi-sovereign polities is British India's Princely States, whose rulers retained some degree of internal legal authority whilst being severely restricted in their external relations. Non-regulation Provinces, such as Kumaon, represent a further but less well known variation of an administrative unit in which "the application of imperial law routinely required its own suspension" (Ogborn 2012: 815).[8]

The unevenly layered domains of sovereignty to which Benton attends largely in terms of jurisdictional politics have been complemented by Ludden's (2011: 143) focus on "the everyday politics of haggling and negotiation" in imperial margins. According to his analytical framework, empire is best understood as a process that operates through patterns of inequality, is driven by negotiations over territoriality, and is characterized by an uneven spread and plural articulation of sovereignty. In order to conceptualize the reproduction and transformation of imperial space

[8] For the emergence of quasi-sovereign places beyond the colonial territory of Britain's Indian empire, see Onley (2009).

Ludden adopts two concepts, fusion and fission, that have a long tradition in anthropological discussions of politics and kinship amongst segmentary lineage societies (Ludden 2011: 139–140). Under fusion, heavy central investments are deployed to tighten the imperial grid and ensure the replication of centralized power and authority in the peripheries. Fission, by contrast, is advanced by frontier activists, such as the commissioners of Non-regulation Provinces, who are relatively independent of central support, command and control. The concepts of fusion and fission thus partly represent versions of Agnew's sovereignty regimes, with the former matching the classic and the latter the imperialist regime. Whilst Ludden (2011: 146–147) points out that these modes often occur within one and the same imperial project, he also highlights that fusion was more prevalent in imperial China whilst Indian empires, including Britain's Indian Empire, remained by and large fissiparous.

In fissiparous empire, elites at higher echelons remain satisfied if the loyalty of local communities in peripheries is secure. Whilst there is always the danger of rebellion and breakaway territories, fission keeps the empire highly flexible and adaptive to changing circumstances. Due to a relatively low level of central investment, the external frontiers of fissiparous imperial expansion are characterized by what Ludden calls 'frontier penury': wealth, tax and tributes move from peripheries toward the center, whilst honor and privilege move from the center toward the margins (Ludden 2011: 138–139).[9] Those who are commissioned to govern such penurious frontiers must therefore generally work with what and whom they find on the ground, which is to a large degree dependent on the activities of preceding regimes. In other words, "[o]ld imperial frontiers [...] move across boundaries of time, creating borderlands among overlapping regimes" that are shaped by plural and often also conflicting sovereign claims (Ludden 2011: 138). It is such multiple claims that are analyzed in what follows by using the example of Kumaon. The region's frontier people, the Bhotiyas, were far from being passive victims or mere recipients of elite imperial plans. On the contrary, they actively negotiated and co-produced the forces through which the Himalayan border region evolved and shifted its shape as an imperial environment. To make this explicit, let us first explore the shape of the Bhotiyas' trans-Himalayan trade within pre-colonial power constellations.

2.2 Trans-Himalayan Trade Within Pre-colonial Power Constellations

Existing scholarly accounts on Kumaon's pre-colonial history are rare and there still remain many discrepancies and open questions. My approach to it is based on the juxtaposition of an oral history I recorded in the present with written colonial

[9] Another important factor in that regard is the movement of various kinds of documents such as letters within imperial space (Ogborn 2007).

source materials. This decidedly eclectic procedure is inspired by studies on social memory.[10] According to Blackburn (2003: 337)

> the idea of social memory reinforces the legitimacy of oral history, or at least places it on a par with other technologies of remembering the past, such as the thin descriptions contained in written sources. In this way, the concept of social memory renders obsolete those tiresome and debilitating dichotomies of myth vs. history, and oral vs. written.

In other words, the concept of social memory reminds us of the fact that societies record, remember and make use of their past through various media besides written texts, for instance through oral memorizations, cultural performances or the establishing of material artifacts (Connerton 1989; Assmann 2011). Borderland scholarship has recently provided new impetus to such approaches, specifically with regard to the interaction between orality and literacy. According to Scott (2009: 220–237) orality, or, to be more precise, the deliberate strategy of language loss, forms part of the Zomian repertoire of state evading strategies. By becoming illiterate, mountain dwellers could have effectively prevented government control, as their lives remained then, quite literally, illegible to civil servants. Taking a less radical stand, Ludden argues that in the imperial margins "archives are typically polyglot, dispersed, obscure and contradictory" (Ludden 2011: 136). The source materials that we find in such archives are often already creolized and the result of multiple translations, a circumstance that, according to Ludden, is typical for a borderland's 'language of power'.

Building on these considerations, the following analysis identifies and elaborates on various thematic connections between both oral and written source materials. Let me start the discussion by introducing an account on the onset of trans-Himalayan trade as contemporary residents of the Darma valley envision it[11]:

> In former times Cyarka Hya ruled over the Darma valley.[12] He was killed soon after the birth of his two sons named Rani Sai and Dasi Sai. Since both of them were still infants at the time of their father's death, it could not yet be decided upon who was the stronger and more powerful one. Through night and day their mothers did everything to raise them, cherishing the hope that one day her son would take on the leading role. But things turned out differently. At around the same time the head of a small chiefdom in the neighboring parts of Tibet had passed away without leaving behind an heir. A lama concluded that one of Cyarka Hya's sons was destined to fill the vacancy and accordingly informed his people: 'We have to identify and kidnap Cyarka Hya's true successor!'
>
> Four warriors, in the guise of mendicants, set off from Tibet to find the two young boys. After long and tiring days of travel they arrived at the doorstep of Cyarka Hya's house in Dantu village at the centre of the Darma valley, begging alms from the widows. In the

[10] The relationship between history and anthropology has influentially been addressed by Sahlins (1985, 2004), who argues that the course of past events needs to be understood in relation to certain cultural structures by which different groups make sense of them. Such structures are transmitted in ritual, myth and cosmology. For a thorough critique of Sahlins' structuralist approach see Obeyesekere (1992).

[11] Versions of this narrative are also given in Leder (2003: 196) and Dhakriyal (2004: chapter 13).

[12] *Hya* means 'elder brother' in the languages of Byans and Chaudans, for which the common word in Darma is *pu*. However, *hya* is also used in ritual and mythological jargon as an honorific for king-like figures as well as divinities.

meantime the four warriors closely watched at the two boys who were playing in the yard. They were instantly affected by their incredible beauty and charm. Both of them were almost identical in their manner of speaking, moving, and behaving as well as in their physical stature and appearance. Since it was impossible to identify the more noble and virtuous one, the warriors took the decision to carry off both Rani Sai and Dasi Sai. They withdrew and waited for the right moment to get in action.

The final kidnapping did not remain undiscovered for long. As soon as the mothers noticed the disappearance of their beloved sons, they chased the Tibetan warriors until they got finally hold of them near Go village, from where the path leads off towards the upper mountain ranges. Although the desperate widows tried everything to wrest their beloved sons from the hands of those men, the situation was hopeless. Alarmed by the tumult residents of Go village hasted toward the place in question, but at their arrival the kidnappers had already disappeared and hid themselves and the young boys in one of the area's numerous mountain caves.

Only at night the warriors stepped out of their retreat, covered the boys' eyes with a piece of cloth and quickly moved northwards until they reached that huge mountain range in front of a place called Widang Padan. While the boys took a rest, the warriors went off the trail to hunt deer. Once the meat was prepared and the boys started eating, the Tibetans gazed at Rani Sai and Dasi Sai with great interest. They determined that Rani Sai paid much more attention to the meat's quality than Dasi Sai, which revealed to them the more noble character of the former. While Dasi Sai was sleeping, the four warriors hastened away with Rani Sai. When they reached Tibet, everyone treated the boy with great care and respect. He received splendid meals including milk, butter and different varieties of meat. Living in a new environment Rani Sai, however, quickly got tired of the local dishes and demanded the Tibetans to bring the food made of gold and silver that he had enjoyed throughout his entire childhood.

Again, the lama ordered the four warriors to the Darma valley in order to obtain these unknown products. What they discovered was the golden flour of *bai* and the silver flour of *palti.*[13] After they had learned the ways of preparing food from these grains, they exchanged a sufficient amount of it for salt and returned back home. Upon their arrival they immediately prepared what Rani Sai had eagerly waited for, such as *syildu* from *bai* and *guthe* from *palti.*[14] The boy remembered the gods and goddesses of his native land and took his meal with great delight. Since then, people say, the barter in grain for salt became a regular affair and soon included further commodities such as wool, animals and borax.

Academic history locates the origin of trans-Himalayan trade in Garhwal and Kumaon within the first millennium B.C. (Joshi and Brown 1987). It is generally accepted that the traditional barter complex evolved around the exchange of grain from India for Tibetan wool, salt and animals. To that extent, the oral history paraphrased above mirrors a much-noticed circumstance, namely "the vulnerability of Tibet's valley economies in terms of grain self-sufficiency" (van Spengen 2000: 98). However, the narrative offers more than a mere statement on the ecological premises of trade. It foregrounds the desire to fill a power vacuum as the major driving force for establishing an exchange of resources that consists of both commodities and personnel. Although the narrative cannot be taken at face value as an historical event, it hints at what Marshall Sahlins (2010: 105) envisages in terms of "mutual

[13] *Palti* (*Fagopyrum esculentum*) and *bai* (*F. tataricum*) are two varieties of buckwheat that the Darmani traditionally cultivate in their upper settlements.

[14] *Syildu* are small pieces of uncooked dough made from buckwheat that are put in a conical shape; *guthe* is a kind of pancake.

relations of determination with neighboring societies". Typically such relations involve strategies of assimilation and differentiation, which are geared both to proximate others and to regional centers of culture, authority and value. A short review of the shifting power constellations in Kumaon helps to elaborate this point and estimate its lasting relevance for later developments.

The disintegration of Katyuri kings' dominance during the twelfth century favored the gradual rise of the Chand dynasty in Kumaon. The latter based their rule on the politico-economic and ritual superiority of high-caste immigrants from the plains, the so-called Thuljats.[15] In other words, the Chands' kingdom bolstered its political clout by inviting high-status plains people to come. The invitees' may have joined this particular constellation, lending it prestige and legitimacy in return for some power of their own. Despite their growing regional influence, the Chands' authority weakened towards the north where new Katyuri offshoots, especially the Raikas of Doti and the Rajwars of Askot had established themselves (Walton 1911: 162–189). The Darma and Johar valley remained tributary to these petty chiefdoms until their direct incorporation into the expanding Kumaon kingdom during the seventeenth century. Things looked different in the valleys of Byans and Chaudans, where the allied royalties of Jumla and Humla exerted control (Walton 1911: 168; Schrader 1988: 102).[16] These ruling houses wielded a lasting influence in the western parts of Nepal and Tibet (*Ngari*) until the rise of the Gorkhas reshuffled the entire socio-political landscape in the central Himalaya during the eighteenth century (Jackson 1976, 1978).[17]

Whilst integrating all Bhotiya valleys for the first time within single polity, the Gorkhas' annexation of Kumaon in 1790 under Amar Singh Tapa fuelled competition amongst the conquered people for a claim to high status within the newly

[15] The traditional social structure in Kumaon and Garhwal exhibits various similarities (Berreman 1963; Sanwal 1976). A widely discussed distinction is that between the Doms, Khasas (also Khasiyas) and Thuljats, with the latter being the smallest but most powerful section. Khasas and Thuljats comprised a common social class, called Bith, whose members segmented into Brahmins and Rajputs. However, the Thuljats always ranked higher as a whole in comparison to any Khasa counterpart. According to Guha (1989a: 12) the overall structure articulates a division "between ruler and ruled" along two basic oppositions: Bith (clean) vs. Dom (unclean) and, within Bith, Thuljat (immigrant) vs. Khasa (indigenous). However, already Srivastava (1966: 187–196) highlighted that the Bhotiyas do not fit in seamlessly with this structure. Under British rule the Bhotiyas from the Johar valley did eventually consider themselves on the same level as immigrant Rajputs (Thuljats), while resident Pahari treated them as equals to the Khasas. Both the Johari and the Khasas classed the Bhotiyas from the Darma, Chaudans and Byans valley with non-Biths, i.e. as Dom, while these latter groups matched themselves as equals to the Johari. The reader should note that after India's independence the usage of these terminologies has become highly controversial and is nowadays perceived as derogatory and avoided (Sax 1991: 65, footnote 26).

[16] The early kingdom in the vicinity of modern Jumla is referred to as Ya-tshe in Tibetan sources (Jackson 1978: 210).

[17] *Ngari* is a general Tibetan place name that is used to designate most of western Tibet. According to Jackson (1976: 39; cf. Francke 1907; Vitali 2003) this region had not been integral to the early Tibetan kingdom until it was brought under central control around 645 A.D. Afterwards it became identified as a conquered land (*mNga' ris*, or 'domain under control'). All major trade marts visited by the Bhotiyas fall within this region.

evolving power structure (Brown 1984: 123–133). This becomes visible, for instance, when comparing the case of the Darma and Johar valley. The residents of these valleys, despite or perhaps even because of their former tributary relation to the same king or *raja*, positioned themselves rather differently in terms of both politico-economic and ritual relations (Srivastava 1966). Most visibly, this took the form of livestock raids and plunder both to meet the oppressive revenue demands raised by the Gorkhas and to outstrip one's neighbor by gaining more influence in the trade marts of Tibet (Moorcroft 1818: 502). Moreover, residents of both valleys eventually appropriated and accentuated prestigious 'foreign' identities in order to demonstrate their supremacy. The migrations of two names are an apt illustration of that process.[18] These are the clan name 'Rawat' in case of the Johar valley and the appellation 'Sai' in case of the Darma valley.

According to Walton (1911: 96; cf. Traill 1992 [1828]: 25) "Rawat means a ruler, and the members of the caste who call themselves by that name appear to be descendents of petty civil officers employed under the Hindu kings". Under British rule – as described later – members of the Rawar clan from upper Johar received unprecedented celebrity as spies surveying Tibet for the colonial government. At this point it is interesting to note the clan's pre-colonial success story, which people associate with a major transformative phase in their valley. The first Rawat is said to have established himself in upper Johar at a time when the people of Kumaon were plagued by the spread of political turmoil and disease.[19] He founded the village Milam that quickly developed into one of Kumaon's most important trading centers (Nüsser 2006).[20] Members of the Rawat clan trace their ancestry from Dham (or Dhamu) Rawat, an immigrant Rajput from Garhwal.[21] While being on pilgrimage in western Tibet, Dham Rawat joined a band of warriors from Ladakh; and together they saved an influential monastic leader from some misery. Rewarded with a lucrative trade permit Dham Rawat completed his pilgrimage before heading home via a route to the east of the mountain Nanda Devi. This route led him directly into the Johar valley where he decided to settle down due to its favorable location. In the course of time more and more 'outsiders' – Rawats and their servants – followed and prospered on the basis of the competitive edge ensured by the clan's profitable agreement for commerce in Tibet.

As Traill (1992 [1832]: 131), the second commissioner of Kumaon, commented: "trade from each Ghat [valley] is confined to some proximate [Tibetan] town,

[18] Here I follow Ramble's (1997: 495; cf. Turin 2002: 253) suggestion that "we may well be dealing not with the wanderings of tribes but the migrations of names".

[19] Mythological accounts of Johar's settlement history are given in Srivastava (1966: 185–186), Leder (2003: 196–197), and Kak (2004: 140–141).

[20] According to a common proverb in Kumaon 'children got lost in the alleys of Milam', which means that it was a big settlement.

[21] It is sometimes also told that there were actually two brothers, one who settled in the Mana-Niti region of Garhwal, the other in upper Johar. Whereas marriage relations between the residents of these valleys have been common, the Johari did usually not intermarry with their eastern neighbors from Byans, Chaudans and Darma (Brown 1984: 63–81). The residents of the latter valleys, however, maintained marriage relations towards Nepal.

beyond which the Bhotias [sic] are prohibited from proceeding without special license obtained from local authorities, the Juwári [Johari] Bhotias alone, in consideration of military services rendered by their ancestors, enjoying an immunity from these regulations".[22] This privileged status in the trans-Himalayan trade circuit might have been one of the reasons that encouraged the colonial administration to establish a rather close relation with the residents of Milam. The Rawats themselves, however, emphasized their alliance to Nanda Devi (Traill 1992 [1832]: 114). The eponymous seat of this goddess, the mountain peak that delimits their valley to the west, evolved as a prevailing symbol of political power. Already one of the earliest Katyuri kings, ruling from Joshimath, "styled himself as the *paramabhakta* or 'greatest devotee' of Nanda Devi in the mid-ninth century" (Sax 2010: 336). Ever since Ajay Pal, the founder of the Shah dynasty who unified Garhwal's petty chiefdoms in the fourteenth century, the goddess Nanda and her natal home (*mait*) in Chandpur became "'keys' to authority over the land and the people" (Sax 1991: 163). It is worth mentioning, for instance, that for the Shah dynasty's greatest rival, the Chands of Kumaon, Nanda Devi was revered as the lineage goddess (Sax 2009: 178).

The reign of Baz Bahadur Chand, king of Kumaon from 1638 to 1778, may have created a significant political impetus for increasing the goddess's significance in Johar. In the course of various military campaigns the *raja* defeated the Garhwalis, took the idol of Nanda Devi from Garhwal to his royal seat in Almora, and also recruited several Rawats whom he then installed as petty officers in the northern peripheries of the expanding Kumaon kingdom (Walton 1911: 177–178; cf. Srivastava 1966: 183).[23] This seems to tally with Johar's settlement history and further contextualizes local residents' claim to be among the greatest *bhaktas* or devotees of Nanda Devi (Srivastava 1966: 201). Their trade caravans traditionally sought protection from this goddess and her worship became a major ritual event in the upper valley, particularly for the Rawats of Milam (Traill 1992 [1832]: 114;

[22] More than half a century later, the British had by then repeatedly cooperated with members of the Rawar clan, Sherring (1974 [1916]: 349) completed this information as follows: "The Rawat ancestor of the Milamwals [natives of Milam village] obtained permission from the Gartok Garphan [governor] to establish himself in trade and built Milam and Burphu, and received a grant of Chunpal [a type of land grant] from the Huniyas [Tibetans]. The connection with Tibet is still kept up, in that the headman of Milam has a so-called Jagir [another type of land grant] at Khyunglung in Tibet, which entitles him to receive annually as a gift five goats and two rupees worth of butter, and as many beasts of burden or coolies as are necessary for the carriage of his effects, whenever he goes to, or returns from, Missar [also known as Moincer, a settlement and former encampment located along the Gartok-Lhasa route south-west of Mount Kailash] in Tibet". The British distributed similar grants to selected individuals, particularly the Rawats, from the 1870s onwards.

[23] Traill (1992 [1832]: 102) provides the account of a local narrative according to which Baz Bahadur Chand built a more accessible path through Johar for which he paid "a rupee with his own hand, for every cup full of earth brought to the spot". This emphasizes the perceived politico-economic importance of wielding influence in these parts of Kumaon.

Leder 2003: 143).[24] But although the cult of Nanda Devi spread far and wide, "she is by no means worshipped universally" (Sax 2009: 179). Among the residents of the Darma, Chaudans and Byans valleys, for instance, her cult is at least nowadays of considerably less importance. A re-consideration of the aforementioned narrative about the origins of trade in the Darma valley points towards one possible explanation for this circumstance.

In several respects the names of the story's two main characters are remarkable. The surnames Rani and Dasi may even sound odd to the reader affluent in modern standard Hindi: *rani* means 'queen' and *dasi* refers to a female slave, servant or concubine. Have the storytellers fooled away and the two sons are in reality daughters? Or, are the names meant to indicate the boys' maternal descent? No matter what, this linguistic irony appears only to contemporary Hindi-speaking people. In the Darma language *ra-* and *da-* are the stems of frequently used verbs, namely 'to come' in the former and 'to give' in the latter case. Whilst Rani Sai might thus be interpreted as the one who *comes* (namely to Tibet), Dasi Sai, as detailed in the following section, can be viewed as the one who *gives* (namely skills and knowledge). The more important issue is, however, the appearance of the appellation Sai (also Se) that follows both names.[25]

The word 'Sai' prevails in the Tibeto-Burman language family, particularly but not exclusively in its Bodic division (Turin 2002).[26] While it has both religious and ethnic-political connotations, the interrelation of both dimensions has so far remained unexplained (Ramble 2003: 69). In *rang-lvu*, i.e. the languages spoken in the Byans, Chaudans and Darma valley, the word for 'deity' is 'Sai' and a number of compounds denote specific divinities such as *Syang Sai*.[27] Persons, usually male, who are versed in certain ritual lore and practices, are called *Sai bano mi*. However, Cyarka Hya's sons are neither revered as deities nor depicted as being particularly skilled in ritual affairs.[28] Rani Sai comes to Tibet as an outsider, who bestows local people with the benefits of a more balanced diet in a kind of *mission civilisatrice*. Dasi Sai plays a similar role for the Darmani: he learnt the Tibetan art of preparing black pudding and encouraged his fellows to learn it too, for which a constant

---

[24] Nanda Devi also had a stand in the region's colonial history. Kumaon's second commissioner, G.W. Traill, is reported to have removed the temple of the goddess in Almora, after which he became stuck with snow-blindness when visiting the slopes of the mountain Nanda Devi. Confronted with this "sign of displeasure" he built her a new temple and became relieved from the curse in due time (Walton 1911: 84). Thereafter, the British 'revered' Nanda Devi as the "highest mountain in the British Empire" (Walton 1911: 228).

[25] Some have understood Sai as the Nepali *sai*, meaning 'one hundred' (Ramble 2003: 36, footnote 10).

[26] Although the issue is not conclusively clarified, the language spoken by the Darmani is considered to fall under the Bodic division of the Tibeto-Burman language family (Willis 2007: 275).

[27] 'Syang' means 'great' or 'major'. Accordingly *Syang Sai* is often depicted as Mahadev or, more particularly, as the Hindu god Shiva (Bergmann et al. 2011: 119).

[28] However, Dhakriyal (2004: 270) compares Cyarka Hya to the divine king and lord of wealth Kubera, who is a prominent figure in Hindu mythology and also related to the Buddhist character of Vaisravana.

supply of salt was then required. In that manner the brothers link complementary ecological niches, one suitable for grain cultivation the other known for its salt deposits and pastoral rangelands. Through the brothers' forced physical separation the Darmani were placed in a position where they could maintain an economically beneficial alliance with residents of the Tibetan plateau. This observation brings us to the second, ethno-political connotation of the word 'Sai'.[29]

Of interest is that 'Sai' and etymologically related cognates are frequently used as internal and external ethnic ascriptions in a number of Himalayan languages. Documented examples include Bodic-speaking communities in central and western Nepal (for example the Thakali, Thangmi and Gurung), Newari speakers in the Kathmandu valley as well as Rai-Kiranti speakers in eastern Nepal (Turin 2002). Historically the term's coinage as an ethnic ascription seems to be connected to an ancient polity that is located in the upper Kali Gandaki gorge and known as *Se-rib* (Jackson 1978; Ramble 2003). Gradually, however, it evolved as a kind of "pan-Himalayan descriptive ethnonym" for people of Tibetan appearance (Turin 2002: 260). It may even be speculated that 'Sai' has a similar status in the Tibeto-Burman language family to that of 'Bhotiya' within the Indo-European.[30]

The application of 'Sai' as a descriptive ethnonym sheds some new light on the characters of Rani Sai and Dasi Sai. In common parlance the Darmani employ 'Sai' neither self-referentially nor for neighboring ethnic groups. Most frequently it is used for naming specific divinities. However, what stands out is that in mythological and narrative contexts the term is also adopted for human beings, as in the example under discussion. In some documented instances these human characters are explicitly associated with former polities in Nepal.[31] Although the precise nature of the link between the word 'Sai' as an ethnonym in *rang-lvu* and its occurrence in a variety of languages spoken in Nepal must remain unresolved here, the 'Sai-connection' (Turin 2002: 258–260) draws attention to the importance of west-east connections that has often been neglected in scholarly accounts of the relational fields within which the trans-Himalayan trade has evolved.

The Darmani maintained much stronger ties, including through marriage, toward their eastern than toward their western neighbors (Brown 1987: 63). Historical records suggest that the Byansi and Chaudansi were over a long period of time tributary to the once powerful royalty in Jumla, whereas Kumaoni kings are said to

---

[29] The reader may note that 'Se' is also the name of a proto-clan or tribe. It is supposed that this tribe, whose name is variously mentioned in old Tibetan scriptures, has once settled in eastern Tibet (Ramble 2003: 70–71). However, since there is no evidence that the Kumaoni Bhotiyas have ever linked themselves with this tribe I will not pursue this line of inquiry further.

[30] Ramble (1993: 17) supports such a view by saying that "'Bhotey' is comparable to the Newar word *Sain*, which has become a derogatory expression for ethnic Tibetans". The complexity of ethnic ascriptions and identity claims among Buddhist Bhotiya groups in Nepal is thoroughly analyzed in Ramble (1997).

[31] Trivedi (1991: 201–210), for instance, transcribed and translated a story from Byans entitled 'Sai-women from Jumla' (*Jumla ci Sai-sya*) who are thought to have once intermarried in Byans. 'Sya' is a widely used feminine gender marker in *rang-lvu*. Folk etymology also relates this marker to the homophonous word for flesh/ meat (Willis 2007: 175).

have controlled the Darma valley.[32] However, colonial sources also report that the Darmani paid dues to various ruling powers in western Nepal (Political Department 1896a: 30). Their tributary relations reached as far as Mustang and thus point to an area that is intimately tied to the genesis of the term 'Sai' in its ethno-political sense (Ramble 1998: 124; cf. Turin 2002: 260). While the north-south axis was frequented for doing businesses, the west-east axis constituted an important dimension for forging political alliances. These alliances have bestowed involved traders with honors (such as exclusive trade permits) and protection (for example, from looters). In my view, it is this kind of alliance relationship that the names Rani Sai and Dasi Sai convey behind the façade of irony.

Sahlins (2010: 218–219) adopts the term 'symmetrical schismogenesis' to describe a prevailing strategy in local competitions over cultural, political and economic resources, namely to engage "powers above and beyond the society and its customary bases of authority".[33] On the one hand we have seen this in Johar. The Rawats charged and performed their high-caste 'Rajput-charisma' (Brown 1984: 29–34) as devotees of the goddess Nanda Devi by tying on the Katyuri-Shah-Chand connection. The Darmani, on the other hand, entwined divine potency with real-life politics along a rather different set of dependencies that point towards neighboring polities in Nepal whose rulers may not even have known of that goddess. It remains a moot point to ask whether valley residents actually submitted to those who brought their political influence to bear in these areas, since alliances with more than one ruler, not least in the interest of their ongoing trade businesses, are reasonable to assume (Srivastava 1966: 184). It is therefore important to note that the Bhotiyas in all valleys were *neither* part of one centralized *nor* of one decentralized state or state-like formation, but rather active agents in "a series of polities existing in a continuous and linked field" (McGranaham 2003: 54; cf. Samuel 1993: 3, 586, footnote 1).

Similar to the kind of 'galactic polities' described by Tambiah (1977; cf. Inden 1981, 2000: chapter 6), the Kumaon Himalaya was pervaded by various shades of political power and authority. This meant that relationships of alliance and allegiance

---

[32] Confirming the impact of Jumla, Traill (1992 [1832]: 129) reported that in 1823 all the villages of Byans "were subjected to a forced contribution by a party of Játs, from Jumla, on some antiquated claim of tribute". Today's residents of Gunji village in Byans regularly perform a ritual that depicts their disengagement from the Jumla royalty (Leder 2003: 187–188). They build an effigy of a king referred to as Jumli Hya that is filled with the innards of a sacrificed goat and transported to a place on the Indo-Nepalese border. All male participants attack and ultimately destroy the image before marking their foreheads with the victim's blood and returning proudly to their village.

[33] The term 'schismogenesis' literally means 'creation of divisions'. Sahlins has borrowed this terminology from Gregory Bateson (1935, 1958) who distinguishes between complementary and symmetrical forms. The former works on the basis of structural oppositions between unequal partners, the latter as a kind of competition between categorical equals. The local appropriation of 'foreign' identities is a strategy that characterizes symmetrical schismogenesis and often results in a 'politics of transcendence', i.e. a quest for divine power and cosmic legitimization (Sahlins 2010: 218–219).

remained in a constant state of negotiation (Burghart 1992).[34] Within such a borderland of confluent territories and overlapping sovereignties the Bhotiyas appropriated the culturally coded ranks of status, honor and privilege of distant others (e.g. regional rulers) in order to outdo proximate rivals, such as the residents of a neighboring valley. The British added new challenges and opportunities to these tension-filled negotiations when they fully joined the game at the beginning of the nineteenth century. It was William Moorcroft who fired the first shot when he set off on an unauthorized journey to western Tibet in 1812.

2.3 Moorcroft's Journey, or: The Discovery of Black Pudding

William Moorcroft, Britain's first academically trained veterinary surgeon, followed an appointment of the East India Company as superintendent of the Bengal stud at Pusa from 1808 onwards (Alder 1980: 175).[35] His main objective was the acquisition of a suitable breeding stock for the not yet well-developed British cavalry in India. This served him as justification for a series of swashbuckling adventures in and beyond the largely unknown lands of the Hindukush-Karakoram-Himalaya mountain complex. The first of his journeys, conducted in 1812, led him via Kumaon and Garhwal to the sacred Lake Manasarovar in western Tibet. The endeavor was demanding in every sense of the word, not least since neither his superiors nor the Gorkha and Tibetan authorities had approved it! In order to arouse as little attention as possible, Moorcroft exploited an institutionalized feature of trans-Himalayan traffic, namely its close relation to pilgrimage (van Spengen 1998).

Adopting a strategy that bears resemblance to the disguised Tibetan warriors in the narrative of the Darmani, Moorcroft and his fellow campaigners took the identity of Hindu trading pilgrims or *gosains*. George Bogle – who had relied on the services of *gosain* Puringir during his 1774 mission to Tibet's second largest city Shigatse – observed that these holy men conducted their journeys not only for gaining spiritual merit but also for economic profit (Markham 1875: 313–315).[36] The *gosains* were always great traders and very active, for instance, during such major festivals as the *kumbh melas* (Cohn 1964). Moorcroft – alias *gosain*

---

[34] The concept of 'galactic polity' is used by Tambiah to characterize a pattern of state formation in which the political influence of a capital center expands through the creation of various sub-centers in a planetary like network. While these networks are characterized by constant change, or systemic pulsation, each sub-center always replicates or imitates the larger order. Lewis (1994: 28) argues that the Himalayan border region was shaped by such kin/state galactic dependencies.

[35] Pusa is located in north-central Bihar. The British Government acquired it in the late eighteenth century for establishing a horse breeding estate.

[36] Bogle also reported that these pilgrims were appreciated by the Tibetan merchants for being capable to adjust to the hot climate of Bengal, where "their [own] people would die" and thus could not go (Lamb 2002: 260). For a more detailed discussion of these Hindu trading pilgrims and their relationship to the British see Clarke (1998) and Bray (2009).

Mayapoori – also gambled for high economic stakes, which "boiled down to breaking to British benefit the monopoly of the so-called cashmere wool shared among Tibet, Ladakh and Kashmir" (Marczell 1997: 166; cf. Pemble 2008: chapter 3). The British East India Company had long seized control of this lucrative commodity and contemplated possible ways to channel it directly to British manufacturers as well as to export live cashmere goats to breeders in England (Alder 1980: 185–186).[37]

Moorcroft left British territory on 7th May 1812 and followed a little-frequented route via Kumaon to Joshimath, and from this former seat of the Katyuri dynasty along the Niti *ghat* (valley) toward western Tibet. He was accompanied by two like-minded companions, captain Hyder Hearsey and his Afghan attendant Gholam Hyder Khan, a large number of porters as well as two Kumaoni *pandits*, or Brahmin scholars, who assisted in conducting geographical enquiries, especially the route survey (Alder 1985: 132).[38] The information-gathering technique on which Moorcroft relied captures very nicely how the ideological basis of imperial expansion, including its prevailing 'Enlightenment mission' (McKay 2003: 63), was actually put to work and became adapted to local circumstances.[39] Under the pretext of enhancing the human condition in the non-European world, this mission paired a passionate thirst for scientific knowledge and enquiry with strategies of economic exploitation. In his published journal, Moorcroft (1818: 382) described his general approach as follows:

> As the road to Jóshí-Mat'h [Joshimath] is known by the surveys [...] I have not been very particular in describing it: but, as the road to Nítí and onwards is new ground to *Europeans*, I shall follow it with more exactness. The principal part of the minutes of our route is taken from the note book of Mr. Hearsey, who carried the compass and brought up the rear, accompanied by Harkh Dev, and who engaged on setting out, to execute this part. Harkh Dev Pandit was directed to stride the whole of the road at paces equal to 4 feet each.

---

[37] The so-called cashmere goat is a central Asian variety of the common mountain goat, *Capra hircus*. However, as Irwin (1973: 4–5) remarks, "[a]lthough goats were the main producers of shawl-wool, a similar fleece was derived from wild Himalayan mountain sheep such as the Shapo (*Ovis orientalis vignei*), the Argali (*Ovis ammon*), the Bharal (*Pseudois nayaur*), and the Himalayan Ibex (*Capra ibex*)". The fine undercoat of these animals is used for textiles and clothing articles.

[38] For the purpose of gathering information British officials relied on the employee of natives quite early. Waller (1990: 22) mentions a sepoy officer who gathered information on the region between Bengal and the Deccan in 1774 as the first recorded example. Subsequently, most recruits were *munshis* (native teachers or secretaries, often Muslims, who also played a crucial role in the administrative system of the Mughal Empire) or, as in the present example, *pandits*. The English loanword 'pundit', which the British later adopted as an appellation for native surveyors, derives from Sanskrit *pandita*, which refers to a male Brahmin acting as "scholar, teacher, adviser, spiritual master, specialist and legal expert" (Michaels 2001: 3). Especially his skills in memorizing and reciting the ancient Vedic scriptures and in mastering special traditions of writing – the *sastras* – are distinguishing skills. The two Kumaoni *pandits* who accompanied Moorcroft were Harbalam and his nephew Harkh Dev. The former had already visited Tibet before, namely as a guide of Webb and Raper on an expedition to trace the source of the Ganges in 1808 (Raper 1812). Captain Hearsey had also taken part in this expedition (Pearse 1905: 180) and subsequently introduced Harbalam to Moorcroft (Alder 1985: 130).

[39] A detailed analysis of the entanglement between enlightenment, imperialism and the British encounter with Tibet based on the endeavors of Bogle and Younghusband is given in Stewart (2009).

We can see here a prototype procedure of colonial knowledge production through a chain of native informants and cultural intermediaries (Cohen 1985; Dirks 1993; Bayly 1996). The Kumaoni *pandits*, educated natives with prior travel experience in Tibet, reported their counted paces to captain Hearsey; the latter, an Anglo-Indian soldier with direct access to the 'other' or non-British culture, transferred the numbers into his notebook and enriched them with technically more advanced measurements. From these raw materials Moorcroft crafted a master narrative – embellished with personal experiences as well as with botanical, zoological, geomorphologic, regional-cultural and politico-economic remarks – that was then prepared for publication by Henry Colebrook, president of the Asiatic Society of Bengal.[40] But how did this many-faced advance guard of imperial expansion identify those who were not, or not yet, absorbed into its fabric? Moorcroft's (1818: 423) first intimate encounter with Bhotiya traders on the Tibetan plateau is telling, and reads as follows:

> At 6,845 paces arrive at our encamping ground, which is bad, as there is only one small spring, from which the water is taken by a small ladle as fast as the hole fills, and this is very slowly. [...] As the water was not in sufficient quantity [...], another well was dug, which gave what was required for cooking [...]. On quitting the spring in the way of our encampment, we saw some of the *Bampo Marchas* [Bhotiyas from the Niti valley] sitting by their loads, with a dead sheep lying on the ground in its fleece, but having the entrails taken out; on a dish lay some of the intestines cut in length like black puddings, and I was led by this to inquire what they were and how made; and from the intimation which I received, was a little surprised to find the borderers of *Bootan* [Tibet], well acquainted with the art of making black puddings. The carcass of the sheep was afterwards, I understood, roasted whole, by being frequently turned in a fire of furze roots.

The Bhotiyas' art of slaughtering and butchering an animal left a lasting impression on Moorcroft: being far-off from British territory he likely did not expect to find meat products that are also sold at London's Smithfield market.[41] He was particularly captivated by the Bhotiya skill of preparing black pudding, from which grew a feeling of familiarity and empathy towards these complete strangers. He, of course, had ulterior motives in cultivating a relationship with them, this being primarily to divert the trans-Himalayan goods traffic to the benefits of British investors who aspired to extend their influence toward the politically significant centers of Tibet and Central Asia (Moorcroft and Trebeck 1976 [1837]: xvii; Lamb 1986: 40). Strikingly, however, Moorcroft's portrayal of his first encounter with these Bhotiya traders also ties in well with an indigenous sense of locality and identity, as the continuation of the Darmanis' story, which tells about the whereabouts of Dasi Sai, elucidates:

> Dasi Sai woke up when his brother had long disappeared. Being alone in some unknown place, he immediately felt hopelessly trapped. After all, his eyes had been covered with a piece of cloth on his way there and he was not able to find a secure path back.

[40] Henry Colebrook added a rather critical footnote to the above quoted passage in which he raised doubt about the accuracy and practicability of taking the rather unnatural paces of 4 ft or 1.2 m when traversing a rough mountain landscape. However, the method was not flawed but rather misrepresented in Moorcroft's notes. It simply withheld that the Pundit only counted his right foot alone with a single ordinary stride of 2 ft or 0.6 m (Waller 1990: 23).

[41] Smithfield is London's oldest livestock, meat and wholesale market.

Fortunately, Dasi Sai could brace himself with some of the leftover food from the previous day. While eating he recollected how the Tibetans had dressed the animal and prepared its meat. With nothing else to do, he began to sketch its details on the ground: the animal body's interior and how to clean and process its different parts. After some days a small group of men from Hula village discovered Dasi Sai. The men had gone hunting and eventually noticed the boy who seemed to play a rather strange game alone on the ground.

Having reached the child they enquired: 'Son, what are you doing here at this lonesome place? How on earth did you come here? And what kind of game are you playing on the ground?' Dasi Sai replied: 'Some time ago four strangers had kidnapped my brother and me from Dantu village. When we arrived here, they killed an animal and prepared plenty of meat for us. Exhausted from the long and difficult journey, we immediately felt asleep after our meal. But when I woke up the next morning my brother and the four men were gone. Since I had no idea which direction to follow and since there was also some leftover food, I decided to stay and wait. I am Cyarka Hya's son, please bring me back to my beloved mother'. After telling his story Dasi Sai started to weep, but the surrounding men replied: 'Stop weeping, we will take you home! Keep quiet! But first tell us what you have drawn in the soil all over the place?'

The child explained: 'The picture shows how these men slaughtered the wild animal and prepared its meat. Look, they cut off the dead animals stomach and small intestine, which were carefully cleaned with fresh water. The animal's blood was mixed with salt and poured into the intestine. Some other parts of the animal's innards were chopped and filled into the large intestine. Afterwards both types of meat were put into boiling water, while the carcass was roasted over the open fire. The boiled meat is called *jyama* [black pudding] and *arjya* [a type of cooked sausage from the large intestine]. You can see its preparation on the ground'.

Dasi Sai's character had been molded by his exposure to an unknown environment. Moorcroft too, on his trek up to the Tibetan plateau, became more and more estranged from his accustomed surroundings (Moorcroft 1818: 422–423). Yet, for both Dasi Sai and Moorcroft the 'discovery' of black pudding turned out as a most defining experience. And this was for more than culinary reasons! After having learnt how to prepare this dish, of which salt was a vital ingredient, Dasi Sai is said to have fostered the economic fortunes and prosperity of the valley's residents by urging them to keep alive the trade in grain for salt. Moorcroft, after having witnessed how the Bhotiyas cooked a delicacy familiar to the British palate, was even more impressed by their sense of business and cooperative attitude.

Worth mentioning in this regard are the brothers Deb and Bir Singh Rawat from the Johar valley who negotiated with the Tibetan authorities in Gartok on behalf of Moorcroft; permitting him to travel into Tibet even after his disguise as a pilgrim had been disclosed. Despite the Tibetan prohibition on the sale of cashmere goats and wool to all but Ladakhi and Kashmiri traders (Moorcroft 1818: 490–491), the Rawat brothers could not resist in selling these to a *feringi* or European foreigner.[42] This fortuitous encounter could have been the seed that brought the Bhotiya trade to the forefront of British strategic calculations. Following his return to British territory,

[42] Even the governor of Gartok had sold these commodities to Moorcroft, a transaction for which the former was "later punished by the authorities at Lhasa (Alder 1980: 188, footnote 61). Unfortunately the wool was impossible to clean after the long journey across the high passes into British territory and the exported live animals either died on the way or soon after they had been transferred to Scotland (Marczell 1997: 167).

Moorcroft propounded the need to support the Bhotiya traders and secure the region as a matter of political urgency and responsibility.

On the one hand Moorcroft had embarked on his journey in the knowledge of Napoleon's alliance with Tsar Alexander. When hearing about Russian trade caravans on the Tibetan plateau he regarded them as forerunners of a secret military campaign, entertaining the suspicion of an imminent Franco-Russian invasion across the unattended high and transversal valleys inhabited by the Bhotiyas (Moorcroft 1818: 453). However, in 1812 Napoleon's army was not on its way to India but to Moscow and thereby considerably weakened. Under these circumstances Moorcroft's strategic bombshell, although still appreciated by British militaries, did not provoke the turbulences that it might otherwise certainly have (Alder 1985: 173–175). The idea of a Russian threat should yet evolve as a leitmotif of British frontier policy at a later time.

On the other hand Moorcroft never tired of emphasizing the negative impact of the Gorkhas, whom he depicted as doing "nothing to ameliorate the state of the country [Kumaon], or to increase the happiness of its subjects" (Moorcroft 1818: 393). This phrase became canonical, was constantly reproduced, and used by the British as a justification for colonization. Moorcroft's positive experiences with the Bhotiyas aroused the interest of British investors who sought to benefit from their trans-Himalayan trade. In the interim, however, his knowledge also proved useful for militarized British advances. Only 2 years after the veterinary surgeon had returned from western Tibet, the governor general waged war on the Gorkhas (Pemble 2008: 130–131).[43]

When the East India Company annexed Kumaon in 1815 the Bhotiyas' trade was limited to the barter of locally needed products, such as grain, salt, cloth and wool. Ladakhi traders monopolized the long distance trade in cashmere (or pashmina) wool used for the weaving of fine textiles such as shawls. A primary aim of the British administration in Kumaon was to divert this lucrative trade from Tibet directly into British territory (Rizvi 1999). Kumaon's status as a Non-regulation Province gave the British the flexibility to accomplish this. Its powerful commissioners usually remained in office for long periods so that their decision making would adapt optimally to local conditions. Assisting these commissioners were locally recruited *patwaris*, revenue officers who had a reach to the remote and inaccessible imperial margins. Influential Bhotiya traders were identified as perfect candidates for this office, as they were well connected to the resource rich Tibetan plateau. The British relied on them as political 'ambassadors' and a bridge to the Tibetan authorities. This strategy became more clearly apparent when the Dogras, a ruling dynasty from Jammu that was tributary to the Sikh kingdom, invaded western Tibet under the leadership of General Zorawar Singh in 1841.

---

[43] Moorcroft's timely piece of reconnaissance as well as his contacts across the frontier proved crucial during the Anglo-Nepalese war, as it becomes apparent from the official correspondence between leading British militaries and government officials (EIC 1983 [1824]). Ultimately, Moorcroft also influenced Hasting's decision to permanently annex Kumaon once the British forces had pushed back the Gorkhas beyond the Kali River (Alder 1985: 172–173; Hastings 1858: 1251).

2.4 Securing Traders' Loyalty in a Non-Regulation Province

Due to this conflict the trans-Himalayan trade was severely affected, and the Bhotiya 'ambassadors' were supported so that they could further British ambitions and deliver intelligence. At that time the Company was already at war with China (First Opium War, 1839–1842) and Afghanistan (Auckland's Folly, 1839–1842). It was feared that the Sikhs would use the Dogras' encroachment on the plateau to foster an alliance with the Nepalese and thus endanger British colonial rule in Kumaon (Huttenback 1961). It is thus hardly surprising that the then commissioner of Kumaon – G.T. Lushington – played a major part in handling the extended role for the Bhotiyas. What is striking, however, is that scholars have consistently ignored his substantial influence on later developments (Tolia 2009: 99–136). Lushington respected the claims and interests of resident communities, especially of the Bhotiyas, and in return expected their allegiance in safeguarding the interests of British investors. When several of the Tibetan marts visited by the Bhotiyas were closed due to the Dogras' military action in 1841, Lushington quickly contrived a strategy that would serve both the Company's and the affected British subjects' requirements:

> It appears to me that these somewhat savage yet well disposed subjects of our government are in every point of view entitled to the above amount of indulgence [referring to a remission of revenues in the Bhotiya valleys] – if not more. [...] It is not however on fiscal as on political grounds that I think these remissions should be made. These people inhabit and occupy the tracts in which the passes into Kumaon from Thibit [sic] are situated. It is therefore of the highest importance that the good feeling, which now exists on their part towards our Government should be strengthened and served. In another point of view too, I think these remissions will be politically advantageous. The fact of their having been allowed will be soon communicated by our traders to the Taklakote [Taklakot, a trade mart in western Tibet, see Fig. 2.1] authorities and tend more than any other measure I know of to convince the latter that the Sikh invasion of their country was neither instigated nor countenanced by our Government (Kumaon Division 1842a).

The Bhotiyas, as 'somewhat savage yet well-disposed subjects' of the British government, were meant to be relieved – at least temporarily – from paying further revenues.[44] On the one hand, such a measure was thought to facilitate a "free and unshackled trade" that the Company was eager to achieve (Kumaon Division 1843a; cf. Brown 1984: 90). On the other hand, a generous fiscal policy was aimed at nurturing the Bhotiyas' 'good feeling' toward the British government. Underpinned by a firm foundation of political loyalty, it was expected that local traders would communicate the benefits of British rule to the authorities in western Tibet with a minimum of assistance or financial burden.

Only a few months later the situation had apparently improved. Lushington, however, remained uncertain about whether his strategy had really paid off or if the

[44] Since the onset of their rule in Kumaon the British levied relatively low taxes on the Bhotiyas – both in comparison to other groups in the region and to the sums collected by the former ruling powers (Traill 1992 [1832]; Batten 1851: 258–356).

situation had eased for entirely different reasons. Based on reports he received from a Bhotiya *patwari* who had returned from the trade marts in western Tibet, the situation was assessed as follows:

> It appears that the Lhassa [sic] authorities entertain none but friendly feelings towards the British Government and its subjects [the Bhotiyas] and in proof of this (supposed) friendly disposition, the re-establishment of the trade between our Bhotiyas and their subjects on the old footing is adduced by our informants. Perhaps the peaceable reestablishment of the trade proceeds more from the conviction of its absolute necessity than from any especial feelings of friendship, for it surely cannot be supposed that the Lhassa authorities are ignorant of our rupture with China [...]. (Kumaon Division 1842b)

'Feelings of friendship' may indeed have hardly played any role. The armed conflict had led to a shortage of grains in western Tibet and the need for supplies encouraged a swift resumption of trade 'on the old footing', at least in respect of some of the merchandise transported by the Bhotiyas (Joshi and Brown 1987: 315). But things had not really settled down yet, as subsequent reports from the *patwaris* attested. The armed conflict not only boiled up again but also impinged on an item that was of high value and of particular interest to British investors, namely shawl-wool.[45] Although the main commercial route for this item then ran through non-British territories in the western Himalaya, with Rampur as its chief mart, it also assumed increasing significance in Kumaon.[46]

In the summer of 1842, Lushington sent the *patwari* from Byans valley on a reconnaissance mission to western Tibet. The *patwari* confirmed that the Dogras were reinforcing a traditional prohibition on the sale of shawl-wool to any but Ladakhi traders (Kumaon Division 1842c). The commissioner seemed, however, less surprised about the stoppage of trade than about the gross profits that the Bhotiyas were reportedly able to pocket. In the margins of the original letter he calculated that the profit from one *maund* of wool was as high as 97 %.[47] However,

[45] The main motive behind the invasion of the independent kingdom of Ladakh by Gulab Singh in 1834 was a monopolistic ambition for the shawl-wool trade. However, the invasion had the unintended consequence of diverting the main commercial route of this commodity toward Rampur, the capital of the former kingdom of Bushahr that is located within today's Indian state of Himachal Pradesh. This development encouraged Gulab Singh "to carry his conquests a stage further and invade the shawl-wool producing areas of western Tibet" (Lamb 1958: 40; cf. Rizvi 1999).

[46] During the Tibet-Ladakh-Mughal war of 1679–1684, the *raja* of Bushahr, Kehri Singh, allied with the Tibetan government in Lhasa. After that war a frinedship treaty was established between Bushahr and Tibet. It was agreed that no transit dues would be levied on trade, which helped to launch an annual trade fair in Rampur (Vasan 2006). The institutionalization of this fair was almost foundational for the Bushahr dynasty and consolidated their political influence. Between 1837 and 1840, just before the Dogras started their military campaign on the plateau, the amount of shawl-wool that was imported from Tibet to Rampur-Bushahr increased over 200 % (Lamb 1958: 40).

[47] The basis for that calculation was the *patwari's* following statement: "The Bians [sic] and Dharma [sic] traders buy at Dharchoola [Dharchula, the major winter settlement and trade depot of the people from Byans, Chaudans and Darma] from the Dhotee [Doti] people 20 Bhelees, or round balls of coarse sugar for 1 rupee – for each of these Bhelees the Taklakote [sic] traders used to allow them the value of 8 annas worth of shawl-wool, the price of the article at Taklakot being 25 rupees per maund – but when conveyed to the Bagiser [Bageshwar, an important trade mart in Kumaon] fair it sells for 50 Rupees per maund. [...]" (Kumaon Division 1842c). Sixteen *annas*

later it became clear that the expected economic losses for the British were of minor significance in comparison to the potential political damages. It was feared that the Dogras might gradually restrict the trade in other important commodities, such as borax and salt, which had the potential to weaken the British position in the trans-Himalayan trade circuit (Kumaon Division 1842d).

Lushington understood the danger of the Company's loss of influence to these native rulers. He thus urged his superiors to demand an official explanation from the Jammu *raja* as to why trade restrictions were being imposed on British subjects. This demand, however, was unnecessary as the commissioner later received information that a combined Tibetan-Chinese army had pushed the Dogra forces back toward Leh (the Ladakhi capital which they had occupied since 1834). J.D. Cunningham, a British observer in this area, urged Lushington to continue safeguarding British interests in the Bhotiya valleys.[48] His prompt was geared particularly toward the well known fact that Tibetan authorities collected dues, including land tax, from traders within British territory. But Lushington, intimately acquainted as he was with local practices and market conditions, rejected the advice to intervene. He understood that such a measure would weaken the Bhotiyas' trading position. In a letter to Cunningham he justified this position:

> I do not consider the present a favourable moment for mooting the point [... since it] would be far better to allow things to remain as they are now, and indeed have been since the establishment of our rule in Kumaon. Nor am I without doubts as to the expediency of hereafter pressing this point on the Lahssa [sic] authorities who have up to the present time shown no disposition to molest our traders and who cannot be expected to renounce a long established privilege without a struggle to maintain it, or at least without having their fears or enmity excited by a demand of this kind. (Kumaon Division 1842e)

The commissioner, instead, preferred to sway Bhotiyas' loyalty toward the British through the maintenance of the tax concessions, which was also meant to deliver a subtle message to the Tibetans on the benefits of British rule. By strengthening the Bhotiyas trade on its 'old footing', i.e. as a measure of convenience rather than strict state control, the commissioner intended to activate a political dialogue with the Tibetans without risking any "loss of dignity", should subsequent developments fail (Kumaon Division 1842e). He even cherished the hope that this dialogue would eventually disrupt the prevailing Tibetan "system of exclusiveness as regards foreign nations" (Kumaon Division 1842f).[49] In other words, the Bhotiyas enjoyed the privilege of paying low taxes to the British whilst having the freedom to follow

equaled 1 Indian Rupee. *Maund* is the anglicized name of a traditional mass unit in South Asia. The British set 1 *maund* equal to 100 English troy pounds (37.32 kg). For the Bhotiyas' profit margins see also Mittal (1986: 216–228).

[48] J.D. Cunningham was posted as the main British observer on the Tibetan frontier in Bushahr (Datta 1969 [1968]). His brother, captain A. Cunningham, was a key player in the boundary commissions that were installed between 1846 and 1847 to define the boundaries of Lahul and Spiti with Tibet (Huttenback 1968).

[49] The reason for Tibet's closed-door policy can be traced back to a decision made in 1792. At that time the Gorkhas had advanced into Tibet and the British refused to assist the Lhasa authorities (Anand 2009: 232).

custom in their interactions with Tibetan traders and state authorities. This policy would remain the *status quo* for over five decades. For the time being, however, the political crisis was not yet fully overcome.

While the Dogras' penetration into western Tibet raised fears amongst the British of a rapprochement between the Sikh and Nepali kings, the subsequent Tibetan recapture of their occupied land brought forth other strategic concerns.[50] British officials then regarded as likely that the Tibetan and Chinese authorities would attribute the whole matter to the instigation of the British East India Company. Such an assumption could indeed have been suggested by the fact that the beaten Sikh and Dogra soldiers received, as a matter of both "policy and humanity", protection in Kumaon (Kumaon Division 1842g). Many soldiers escaped into British territory across the Lipu Lekh pass that leads into the Byans valley (see Fig. 2.1). The Bhotiya *patwari* of that valley was instructed to arrange for the soldiers' safe onward journey to the British executive headquarters in Almora, where they received medical treatment.[51]

In order to avoid a direct confrontation with the Tibetans near Kumaon, which Lushington anticipated would result in a complete standstill of trade, serious thought was given by the British to the idea of providing military support to the Dogras in the more distant Ladakh (Lamb 1958: 41). In the end, however, the British did nothing but wait because the two parties continued to battle for supremacy in that locale (Kumaon Division 1842i). On the 17th September 1842, the Tibetans accepted a peace agreement with the Dogras (Warikoo 1990: 138). The Dogras remained in control of Ladakh and traders on both sides continued business in accordance with former custom. In Kumaon, a favourable vantage point for the flourishing trans-Himalayan trade – particularly in shawl-wool – was established in 1843. In that year Gulab Singh, the ruler of Jammu, formally expressed his gratitude to Lushington "for the attention shown at his place in the year 1841/1842 to his sick and wounded soldiers from Taklakote [sic]" (Kumaon Division 1843b). Three years later, and after the British had won a war against the Sikh kingdom, Gulab Singh and the British government were formally allied with the Treaty of Amritsar.

Soon after, the Ladakhi monopoly over the shawl-wool trade was finally and officially dissolved. A rapid increase in the volume of trade, particularly for wool,

---

[50] The Dogras' main military leader, general Zorawar Singh, was killed in a battle near Taklakot in December 1841. His fellow campaigner, colonel Basti Ram, continued fighting in that locality until January 1842 (Charak 2003 [1983]: 763).

[51] Already during the fights around Taklakot in autumn 1841 the *patwari* from Byans valley forwarded to Lushington an account of a fugitive Sikh soldier named Raza Singh: "No arrangement has as yet taken place between the Chinese and Sikhs. Five fights had occurred in the first of which the Chinese lost about 1,000 or 1,200 men besides prisoners – and in the second the loss on both sides was equal and about 100 prisoners were taken by the Chinese. I was among the prisoners. After disarming me they bound me with ropes and sent me in the direction of Lahssa [sic] but I managed to cut the ropes at night and make good my escape to this place hiding by day in the jungle and travelling at night. The Chinese army amounts to about 12,000 or 13,000 men" (Kumaon Division 1842h). In January 1842 colonel Basti Ram also escaped via the Byans valley to Almora. He and his remaining 240 soldiers received full support and medical treatment due to Lushington's supportive orders (Charak 2003 [1983]: 764).

Table 2.1 Import of wool, borax and salt from Tibet to Kumaon in the nineteenth century

	Gori valley	Darma and Byans valley	Total
1840			
Wool	15	7	22
Borax	9,000	8,000	17,000
Salt	2,000	3,000	5,000
1877–1878 (annual average over 2 years)			
Wool	4,325	479	4,804
Borax	10,061	10,335	20,396
Salt	5,177	8,266	13,443
1898–1901 (annual average over 4 years)			
Wool	2,388	6,479	8,867
Borax	6,808	14,687	21,595
Salt	5,510	13,855	19,365

Measuring unit *maund*, where 1 *maund* equals 37.32 kg

Sources: Annual Report on Foreign Trade of the United Provinces, 1878, V/24/4163, IOL; J.E. Goudge, *Almora: Final Report on the Assessment of the Almora District and the Hill Pattis of the Naini Tal District*, 1903, V/27/314/643, IOL (See Bergmann 2016: 95)

emerged during the second half of the nineteenth century (Huttenback 1961: 488; see Table 2.1). The wool trade through the high valleys of Kumaon increased from approximately 800 kg in 1841 to more than 330,000 kg by 1901 (Goudge 1903). Industrialized wool mills in fast-growing market towns along the Himalayan foothills fuelled this growth in demand (Roy 2003: 257–286). Initially the Bhotiyas sold their goods mostly through middlemen in the Lesser Himalaya. However, they later began to bargain directly with representatives of these mills (Walton 1911: 69). The drainage of the *Bhabar* as well as the subsequent extension of the railway network in the 1880s transformed these towns into the most important market centers of Kumaon (Goudge 1903).

The Bhotiyas' trade expanded rapidly once the Dogras' military campaign in western Tibet ended. With a reduction in the perceived risk of foreign invasion via the Tibetan plateau, the strategic importance of their mountain passes waned, at least for the time being. British administrators then considered it neither necessary to protect Kumaon's northern boundary nor to intervene too much in local affairs. As a result, the Bhotiyas had little to which they could object, especially since their (tax) privileges and relative autonomy in handling trade-related affairs remained by and large unchanged. In summary, it can be said that the British capitalized on Kumaon's exceptional legal status as a Non-regulation Province to secure the Bhotiyas' loyalty and therewith strengthen the empire's position within the trans-Himalayan commercial circuit. They achieved this objective through tax concessions and non-interference in the traditional trade arrangements. In implementing these measures, however, British administrators accepted limits to imperial sovereignty in favor of other interests, namely of channeling economically significant trade products, especially shawl-wool, from Tibet to British territory.

Initially these limits were recognized primarily in terms of the strict prohibition on the entry of British officials into Tibet. When the East India Company relinquished control over the Indian subcontinent to the crown in 1858 this closed-door policy provoked increasing political discontent (Anand 2009: 232). Tibet made a mockery of the empire's standardized instrument of power and control; the cartographic map.[52] As a monastic realm that was largely unknown to Europeans this mountain region avoided being absorbed into the European master image of a state crafted according to "the Westphalian concept of territorial integrity and exclusive sovereign autonomy" (Legg 2009: 101). However, in the course of an intensifying rivalry between the British and the Russian empires for supremacy in Central Asia, the so-called 'Great Game', the blind spot that was Tibet became secretly revealed through clandestine surveys. Recruited and trained by the Great Trigonometrical Survey, two Rawats from Milam village in the Johar valley conducted the first of these surveys in the late 1860s. On the basis of their measured data the British could establish precise cartographic representations of Tibet. The overall mission advanced the territorial structuring of imperial space and, eventually, also served the purpose of distinguishing this monastic realm as a neutral buffer zone between the rivaling superpowers.

2.5 Advancing the Territorial Structuring of Imperial Space

It was the British surveyor T.G. Montgomerie who contrived a strategy for achieving the ambitious goal of exploring the Tibetan plateau.[53] His plan was to recruit those "subjects of the British government, [who] are known to travel freely without molestation in countries far beyond the British frontier [since] they constantly pass to and fro between India and Central Asia, and also between India and Tibet, for trading and other purposes, without exciting any suspicion" (Montgomerie 1868: 129). So what could be more appropriate than to recruit Bhotiyas and rely on their assistance? Their subsequent missions show quite clearly that the British Empire evolved and expanded not as a fixed structure but rather as a malleable and highly dynamic process.

Under the approval of colonel J.T. Walker, then superintendent of the Great Trigonometrical Survey in Dehra Dun, Nain Singh and his older cousin Mani Singh Rawat were chosen to survey Tibet in 1863 (Waller 1990: chapter 2). Both descended from well-off trading families in Milam, which was by then one of the most significant trading hubs in northern Kumaon. Nain Singh was the village-school's headmaster, Mani Singh the valley's *patwari* or revenue officer. Their selection was no coincidence. Mani Singh was the eldest son of Deb Singh, one of the Bhotiya

[52] For detailed analyses of surveying and mapping activities in British India see, for instance, Edney (1997) and Barrow (2003).

[53] During the 1850s Montgomerie established himself as the head of the Kashmir survey and subsequently accomplished additional work in Ladakh (Waller 1990: 18–21).

traders who had supported Moorcroft on his unauthorized journey to western Tibet in 1812. Moreover, both recruits already had an untarnished reputation in assisting British officers and European explorers, such as Richard and Henry Strachey in 1849 as well as the brothers Schlagintweit between 1855 and 1857 (Waller 1990: 40).[54]

In order to keep their sensitive undertakings confidential, the British referred to them by codenames: 'Patwar' in case of Mani Singh and 'Pundit', the scholar, for Nain Singh.[55] Since Nain Singh eventually rendered his service far more successfully, his codename was soon adopted as a generic for many follow-up recruits with him becoming the 'Chief Pundit'. During 2 years of training the cousins mastered both standard procedures of topographical reconnaissance and effective techniques of concealment. In 1865 they were sent on their first mission, which was destined to lead them via Nepal to Lhasa (Waller 1990: chapter 2). Due to problems with Chinese guards monitoring the high passes, the secret explorers decided to split up. While Mani's subsequent performance proved unsuccessful, Nain Singh and his servant bluffed their way through and travelled undisturbed on the Gartok-Lhasa route, first with a Bushahri trading group and then with a Ladakhi caravan.

The Pundit's success stemmed from a cunning concealment of scientific expertise, military calculation and superpower politics. Once again the adoption of an old and widespread form of trans-Himalayan traffic, namely pilgrimage, served as a template for secretly moving the imperial frontier without provoking external security risks, disproportionate expenses, or any other threat to the prestige of British rule or the lives of British officers. Moreover, the ritualistic framework was deemed ideally suited for assembling the right competences, devices and people under a stable working routine.[56] This working routine was suggested by the prevailing perception of Tibet as a monastic realm in which people practice a kind of 'robotic' ritual. As one early observer put it, "the duties of the [Buddhist] priests are mere mechanical movements, or 'manual devotions', set a-going at particular times of the day; the operations of a machine which might just as well be put in motion by wind, water, or stream, as by human hand. A cylinder, like a drum, lined with written papers, is whirled around on its axis [...]" (Anon 1817: 423).

Moorcroft's information-gathering technique, as outlined before, provided an inspiring example for adapting these mechanical movements to rather different purposes. The last mentioned instrument, the Buddhist prayer wheel, proved particularly useful for conducting a clandestine route survey. Montgomerie's recruits learned to handle a modified device that was used to store information on measured distances, altitudes, wind velocity, rain gauge and stream. To complete this account it is worth quoting the ingenious approach as it was published in the

[54] Deb Singh's younger son Kishen was also recruited as a Pundit in 1869 and extremely successful (Mason 1923).

[55] Additionally, the reversed order of initials was used, such as GM for Mani Sing(h), in official reports.

[56] In that sense the case exhibits some similarities to the methods of long distance control on which the Portuguese relied to foster their imperial expansion to India (Law 1986).

Royal Geographical Journal after Nain Singh's successful return (Montgomerie 1868: 139–141):

> During the regular survey of Ladakh, Captain Montgomerie had noticed that the Tibetans always made use of the rosary and prayer wheel, he consequently recommended the Pundit to carry both with him, partly because the character of Buddhist was the most appropriate to assume in Tibet but still more, because it was thought that these ritualistic instruments would (with a little adaptation) form very useful adjuncts in carrying on the route survey.
>
> It was necessary that the Pundit should be able to take his compass bearings unobserved, and also counting his paces, he should not be interrupted by having to answer questions. [...] Whenever people did come up to the Pundit, the sight of his prayer-wheel was generally sufficient to prevent them from addressing him. When he saw anybody approaching, he at once began to whirl his prayer-wheel round, and as all good Buddhists whilst doing that are supposed to be absorbed in religious contemplation, he was seldomly interrupted.
>
> The prayer-wheel consists of a hollow cylindrical copper box, which revolves round a spindle, one end of which forms the handle. The cylinder is turned by means of a piece of copper attached by a string. A slight twist of the band makes the cylinder revolve, and each revolution represents one repetition of the prayer, which is written on a scroll kept inside the cylinder. [...] The one used by the Pundit was an ordinary hand one, but instead of carrying a paper scroll with the usual Buddhist prayer 'Om mani padmi hom', the cylinder had inside it long slips of paper, for the purpose of recording the bearings and number of paces, &c. The top of the cylinder was made loose enough to allow the paper to be taken out when required.
>
> The rosary, which ought to have 108 beads, was made of 100 beads, every tenth bead being much larger than the others. [...] The rosary was carried in the left sleeve; at every hundredth pace a bead was dropped, and each large bead dropped, consequently, represented 1,000 paces. With his prayer-wheel and rosary the Pundit always managed in one way or another to take his bearings and to count his paces.
>
> [...] His observations for latitude were all taken with a large sextant, by Elliot, of 6-inch radius, reading to ten seconds. [...] The Pundit had invested in a wooden bowl, such as is carried at the waist by all Bhotiyas. This bowl is used by the Bhotiyas for drinking purposes; in it they put their water, tea, broth, and spirits, and in it they make their stirabout with dry flour and water, when they see no chance of getting anything better. The Pundit, in addition, found this bowl answer capitally for his quicksilver, as it's deep sides prevented the wind from acting readily on the surface. Quicksilver is a difficult thing to carry, but the Pundit managed to carry his safely nearly all the way to Lhasa, by putting some into a cocoa-nut, and by carrying a reserve in cowrie shells closed with wax.
>
> Reading the sextant at night without exciting remark was by no means easy. At first a common bull's-eye lantern answered capitally, but it was seen and admired by some of the curious officials at the Tadúm monastery, and the Pundit, who said he had brought it for sale, was forced to part with it, in order to avoid suspicion. From Tadúm onwards a common oil wick was the only thing to be got. The wind often prevented the use of it, and, as it was difficult to hide, the Pundit was at some of the smaller places obliged to take his night observation, and then put his instrument carefully by, and not read it till the next morning; but at most places, including all the more important ones, he was able to read his instrument immediately after taking his observations.

The distance travelled by this 'unholy spy' was enormous (Hopkirk 1982: chapter 2). On his first mission alone Nain Singh covered 1,200 miles, i.e. two-and-a-half-million individual paces of equal stride length counted with the help of his rosary (Hopkirk 1982: 35). Later comparisons showed the extreme accuracy of his measurements, which enabled the British to fill many blanks in their outdated maps.

Moreover, these journeys enhanced the information basis on various other issues of interest, such as settlement and land use patterns or the modalities of Tibetan gold production (Montgomerie 1868: 165; 1869). For his rendered services Nain Singh received all honors – including a gold watch from the Society of Geographers of Paris, a gold medal from the Royal Geographical Society and the title of C.I.E. (Companion of the Indian Empire) in 1877 (Waller 1990: 52). As it seems, the Pundit's journeys were valued as an outstanding example of natives' commitment to scientific rigor and political duty in a – at least to the western world – hostile and largely unknown environment. But things weren't quite that simple.

The force of these journeys with regard to British strategic interests stemmed from what Allen (2011: 288) calls a 'topological equation'. According to such equations political influence is brokered not only in terms of extensive but also in terms of intensive reach. The resulting arrangements of power enable "distant actors to make their presence felt, more or less directly, by dissolving, not traversing the gap between 'here and there'" (Allen 2011: 290).[57] In other words, they jolt people's understanding of what is near and what is far. While already Lushington had pursued such an agenda by relying on the Bhotiyas as political 'ambassadors' and a bridge to the Tibetan authorities, it was Montgomerie who took full advantage of its potentials. The co-optation of the Pundits allowed the Great Trigonometrical Survey to register the empire's presence far beyond the reach of bureaucratic command and military control. The strategy was most adequate at that time, since the drive for military-based imperial expansion had slowed down dramatically after the Indian Mutiny (1857–1858) (McKay 2003: 63). During a period of cautious consolidation, which lasted approximately until the 1880s, expensive frontier operations were difficult to enforce.

By adapting a local phenomenon, viz. pilgrimage; Montgomerie devised a quiet and cost-effective strategy to leverage British imperial influence on the Tibetan plateau. With every step made within the great distances travelled by Nain Singh, the greater the value of the information that he channeled to his superiors in the Great Trigonometrical Survey. As a Tibetan-looking and Tibetan-speaking British subject he could accomplish a comprehensive route survey and thus provide colonial cartographers with a solid foundation of measured data for 'flattening' a hitherto impassable mountain region, via cartographic projections, to the map. In return Nain Singh acquired the possibility of receiving wealth, such as an attractive salary and superior entitlements to land, as well as honor through official

[57] Topology is a branch of mathematics that aims to articulate the rules for localizing objects in a variety of coordinate systems. Social scientists increasingly adopt topology-talk for differentiating several kinds of space through which the social subsists (Mol and Law 1994; de Laet and Mol 2000; Law and Mol 2001). The use of topology for the explanation of social phenomena already has a long tradition in anthropology: Leach (1961), a trained engineer and mathematician, invoked it "as the geometry of elastic rubber sheeting" for the study of kinship and Levi-Strauss (1985) relied on the topology of Klein Bottles for interpreting and comparing myths. In geographical writings it is nowadays widely adopted to "disguise the relational complexities that lie 'underneath' spatial forms" (Murdoch 2006: 12).

awards and titles.[58] But Nain Singh also repeatedly put his life on the line to accomplish the tasks assigned to him. With every counted pace the Pundit not only demonstrated his tireless motivation and strength of will in covering physical distance but also cemented his allegiance to Montgomerie and the Great Trigonometrical Survey. In that sense he instantiated a flesh-and-blood machination that both instructed and camouflaged the imperial script over great distances – as Rudyard Kipling's (n.d.: 157) protagonist 'Kim' put it – "far and far into the North, playing the Great Game". If the Tibetans had unmasked his 'true' identity, severe punishment, imprisonment or even death would have been likely consequences.

But despite all personal hardships and risks, the Pundit's activities had, taken by themselves, only a limited political force. As Bhotiya, a native subject from the very margins of the British Empire, he was neither enabled nor allowed to compute the results of his own observations:

> The explorers are taught to make a route survey by taking bearings with a compass and pacing the distances; they are also taught to take meridian altitudes with a sextant to determine latitudes; but they are purposely not taught how to reduce their observations, nor supplied with astronomical tables, in order that they may not be able to fabricate fictitious work. Observations for determining absolute or differential longitudes are beyond their capacities. The resulting latitudes and the co-ordinates of the results are computed in the office of the Superintendent of the Great Trigonometrical Survey on the explorer's return. (Markham 1968 [1878]: 148)

It is, of course, hardly unusual that British colonial administrators tried to keep the natives in their 'place' (Lester 2012). The rationale for not allowing the Pundits to perform any kind of calculation clearly indicates a racial anxiety that was widespread in colonial times. It further reveals how the conquest of territory was gradually transformed by a new science dealing with 'incontrovertible' and 'absolutely distinctive' facts, viz. geography (Holdich 1899: 473; cf. Piper 2002: chapter 2; Raj 2002). Within this emerging discipline the entanglement between knowledge and power seemed particularly pronounced. It was therefore not without reason that the Tibetan authorities in Lhasa reportedly feared geographical maps depicting their monastic realm as much as they did an overt military expansion of British rule (Pundit A.K., quoted in Anand 2009: 234). Maps of an ever-higher quality radically altered *what* the British could demand politically as well as *how* these demands were to be brokered and contested. When a British military force, led by Francis Younghusband, invaded Tibet to drive back the assumed Russian influence in 1903, the information gathered by the Pundits, particularly their maps, proved very useful to the troops in navigating the mountains (Younghusband et al. 1905).[59]

---

[58] Nain Singh received the "grant of a village in Rohilkhand together with a *jumma* [revenue assessment from land] of Rs. 1,000" after his career as a spy and surveyor for the Great Trigonometrical Survey (Waller 1990: 125).

[59] After the Tibetans had been defeated a trade agreement was imposed, which sanctioned the establishment of British trade posts on the plateau (McKay 1992). In western Tibet a native Indian, named Thakur Jai Chand, was based in Gartok as a British trade agent (Sherring 1974 [1916]: 148; McKay 1997: 158–165). The Bhotiyas often complained about the futile character of that institution.

In the course of these political developments also British relations toward the Bhotiyas were in a process of change. The traders might first have noticed this change through the activities of a new and powerful institution, namely the Imperial Forest Department. The department was established in 1864 with the help of German forest managers (Guha 2001). Under the guidance of these foreign experts colonial forestry in India was largely implemented without considering local preferences and needs. Hand in hand with the Great Trigonometrical Survey, the Imperial Forest Department fostered the territorial structuring of imperial space for administrative purposes, security reasons and the provisioning of resources. Although there was a marked tendency amongst influential functionaries within both institutions to ascribe a subordinate role to the knowledge of resident communities, there were occasional exceptions. Montgomerie and others who personally engaged with the Pundits deeply valued the natives' work and achievements (Waller 1990: 124–127). For instance, when Nain Singh was awarded the gold medal of the Royal Geographical Society in 1877, colonel Henry Yule (1877: cxxvi) explicated this in public by saying that the Chief Pundit "is not a topographical automaton, or merely one of a great multitude of native employee's with an average qualification. His observations have added a larger amount of important knowledge to the map of Asia than those of any other living man". Also Dietrich Brandis, the German scientist who guided the Forest Department's setup in India, publicly campaigned for an appreciation of local skills (Guha 2001: 222). In other words, even within these key institutions of Britain's nineteenth-century Indian Empire there seemed to be two minds about how imperial policies should be developed and pursued.

Nevertheless, after the implementation of the Indian Forest Act in 1878, the Bhotiyas found themselves enclosed within complex administrative arrangements, especially when visiting the trade marts and pastures in the Lesser and Outer Himalaya (Bergmann et al. 2011). Although the trans-Himalayan trade in Kumaon was still seen as vitally important to British interests, the Bhotiyas' seasonal journeys with large flocks of sheep and goats were increasingly being defined as maladaptive and harmful to commercial forestry (Forest Department 1905; Dangwal 2009). In other words, within an economically motivated rhetoric of nature conservation the Bhotiyas gradually emerged as a disturbing factor (Rajan 2006). Moreover, it was increasingly considered problematic that "[t]he absence of any control over them has in a way spoilt them and they seem to have very little respect for authority. The wandering life they lead [...] encourages lawless habits" (Political Department 1895a).

This gradual shift in perspective brought to the fore another dimension of the limits to British imperial sovereignty. Whilst maps played a key role as to how British officials imagined their political authority, namely as being defined territorially, it gradually transpired that there existed a great uncertainty as to who was in control over what and where. Already in the earliest periods of British rule in

---

One reason for their complaints might have been that the establishment of such 'listening posts' gradually dismantled their own privileged status as both information carriers and political 'ambassadors'.

Kumaon, G.W. Traill (1825, quoted in Walton 1911: 149), the second commissioner of Kumaon, had discerned that the Bhotiyas remained in "an anomalous state of subjection [to Tibetan rule] which their paramount interests in continuing to be the medium of commercial intercourse [...] will tend to perpetuate". The fact that the Bhotiyas continued paying taxes on land to Tibetan authorities even gained tacit approval under British rule. G.T. Lushington had taken a stand for the retention of this practice in handling the region's complex milieu of conflicting interests and alliances during the 1840s. About 30 years later the settlement officer J. O'Beckett, simply reworded the Tibetan land tax as a trade levy so as not to disturb the flow of commerce (Political Department 1896a: 27).[60] In practice, everything continued pretty much as before, not least since British officers remained virtually absent in these valleys (Political Department 1896a: 33). The confluent territories and overlapping sovereignties to which these arrangements point were only questioned toward the end of the nineteenth century, a time when British administrators made serious efforts to bring this external frontier of fissiparous imperial expansion under the colony's standard grid.

2.6 Claiming Territorial Sovereignty and Negotiating for the Formal Rule of Law

As a Non-regulation Province, Kumaon's legal status had remained rather obscure. For this reason the colonial government dispatched a judicial official, P. Whalley (1991 [1870]), to gather detailed information on the laws and regulations that were in force in 1869. Thereafter, Kumaon was included within the Scheduled District Act, which was passed in 1874. This act "listed those districts that would be exempted from legislation applicable to the rest of [British] Indian territory" (Benton 2008: 612). As a result the Kumaon commissioner had some of his powers restricted, whilst new regulations and acts were extended or declared in force in the region (Mittal 1986: 85–86; Tolia 2009: 285–287). The high valleys inhabited by the Bhotiyas, however, remained largely unaffected by these developments.

This situation was set to change when a high-ranking British military officer, major general G.N. Channer, went on a hunting trip through the upper Byans valley

[60] O'Beckett's overall approach was reproduced as follows: "The settlement officer sums up the position [of the British] by saying in the language of these parts: '*Bakri hamara, Karhaj tumhara*' (The goat is ours, the load is yours). Thus it is evident that Mr. Beckett, while disallowing the claim of Thibet [sic] to collect their demands under the name of land revenue or '*rakam*', admitted them, though under a different designation. He called it trade dues. In his Settlement Report he enters it as '*Jagad*'. His words are: '*Hundes men jagad ke ewaz dete hain*' (This is paid to Thibet in lieu of trade tax)" (Political Department 1896a: 27). Actually the Tibetan revenue system was highly complex and consisted of various tributes, such as two types of *sathal* (land revenue and tolls for crossing the high passes), *chunkal* (trade tax), *naika* (trade tax) and *lakpacha* (measuring tax) (Political Department 1896a: 30–32). By simply classifying them as *Jagad* and thereafter sanctioning its collection, the British had "virtually admitted Thibet's claim" to territorial jurisdiction in the Bhotiya valleys (Political Department 1896a: 33).

in 1884. He personally witnessed how British subjects, the Bhotiyas, paid taxes on British land to the Tibetan authorities (Political Department 1896a: 34). Channer and other military men repeatedly urged the Kumaon administration to once and for all prevent "the malpractices of the Thibetan [sic] Jongpen [governor]" (Political Department 1894a: 25).[61] In consequence, the then commissioner of Kumaon raised the issue of the absurdity of leaving unaddressed a "bantam power [that] virtually claims sovereignty" over British subjects (Political Department 1896b: 20).[62] The whole incident touched a raw nerve and triggered a knee-jerk quest for what Ranajit Guha (1989b: 242) calls "strategies of improvement": government initiatives that would give adequate attention to both the security of the Himalayan border region and the activities of its resident communities. It was regarded as particularly important to tie the Bhotiyas more formally to the colonial state by urging them "to recognise British authority" and to "realize their position as British subjects" (Political Department 1895b: 27).

The colonial government had so far interfered in the local affairs of the Bhotiyas only to ensure that trade would continue on its 'old footing'. But it turned out that this fairly lax policy had created its own frictions. Both British and Tibetan state officials staked their claims over the flourishing trans-Himalayan trade. Whilst the British had been aware of these overlapping sovereign claims for decades, they had continued to imagine their political authority as being defined territorially. However, it gradually transpired that there existed a great uncertainty as to who was in control where. Without the presence of clearly demarcated and secured boundaries, the Tibetans effectively wielded influence within these confluent territories through power relationships of allegiance. These relationships were reflected, for instance, in the taxation scheme that the *jongpen* applied in the Bhotiya valleys. In order to ensure that British sovereignty and territoriality become congruent, a scheme "for placing the administration of that tract [the valleys inhabited by the Bhotiyas] on a proper footing" was thus felt to be required (Political Department 1895c: 31, cf. 1895d). A general problem was that colonial officers had become "out of touch" with local frontier people (Political Department 1896a: 34). Since Lushington's death in 1848 their only information about what was happening in the high valleys was from the *patwaris*, influential Bhotiyas who held their office by right of heredity. While all of them had a good name and reputation, they were largely independent of government instruction and control, a circumstance that became seen as highly problematic.

As an initial step the new office of the *political peshkar*, a civil servant with police powers, was temporarily established in October 1895 (Political Department

[61]According to British perceptions these malpractices further included the application of Tibetan law, the prevention of European and Pahari travelers from entering Tibet, the intentional destruction of roads and rest houses, the cutting of wood, the monopolization of grazing grounds and the general disrespect shown to British subjects and political authorities (Political Department 1894b).

[62]Whilst British administrators described the Tibetan claims in northern Kumaon as 'virtual sovereignty' during the 1890s, the vocabulary later shifted to the mantra of 'suzerainty' (Walton 1911: 149). At the beginning of the twentieth century the latter term was adopted to define Chinese claims over Tibet (Goldstein 1989: 830; Anand 2009).

1895c: 31–33). Kharak Singh Pal was appointed because of his social standing and political influence, and because of his competence in a variety of local languages (Political Department 1896a: 34; cf. Brown 1984: 91–92).[63] He was a close relative of the Rajwar of Askot, the pre-colonial ruler of the region whom the British simultaneously installed as an honorary *munsif* or native judge (Political Department 1895c: 32). Besides acquiring information, the *peshkar* was assigned to counter inappropriate Tibetan demands on and dealings with British subjects. Although he was considered to be doing a good job, the need for European presence and active involvement was soon thought indispensable for handling the situation "in an authoritative way" (Political Department 1896d: 26).[64] For that reason, in summer 1896 the deputy collector from Almora, J. Larkin, was dispatched to inspect the region and consult with the *jongpen* of Taklakot, the major political player on the Tibetan side. He returned with the following account of the local atmosphere:

> I think I may appropriately mention my own experience as illustrative of the bearing of the people [the Bhotiya traders] to the English official and towards the Jongpen. When the latter individual approached my camp [...], the Bhutias [sic] to a man deserted my camp and hastened to meet the Thibetan [sic]. They received him, erected his tents, attended to his wants and comforts as well as to those of his large following, their horses and cattle, all gratuitously. During the two days that our camps lay side by side, the Bhutias were conspicuous by their absence from my camp. This and other signs showed me that while our Bhutia subjects were cringingly servile to the Thibetan, they were coolly indifferent to the English official. All this is no doubt due to our supineness in the past as compared with the rigorous rule of Thibet. If our District Officers had visited those parts and looked into the matters these things would be different. (Political Department 1896a: 34)[65]

[63] British militaries had originally proposed to install more than one official, "so that they would be a check upon one another" (Political Department 1894a: 26). Whilst this was never realized, three support bases were established for the *peshkar*, namely in the settlements Garbyang (Byans valley), Go (Darma valley) and Sosa (Chaudans valley) (Political Department 1895e: 5). In the Johar valley, where the conflict was not as pronounced, Dalip Singh Rawat, a Bhotiya who had taken up a career as a civil servant in Almora, was temporarily employed for that office (Political Department 1896c: 23).

[64] The only Western observers who were present in these tracts were missionaries of the U.S.-based Methodist Episcopal Church. They had established themselves in Dharchula under the supervision of Mrs. Sheldon, M.D., Mrs. Browne, and Dr. Wilson in 1883. Besides their missionary work they provided basic medical and educational facilities to the local people (Browne n.d.). As for other missionary stations along the Indo-Tibetan frontier, their overall intention was to enter Tibet (Pande and Bergmann 2009).

[65] Landor (1899: 8) – a British explorer who extensively travelled through the region at that time – described Larkin as one of those rare officials "who had any knowledge at all of the northeast of Kumaon". In his travelogue he provides a slightly different account of the encounter between the deputy collector and the *jongpen*: "Trembling with fear and bending low to the ground, the Tibetans, with abject servility, entered the tent of our British envoy. The account of the interview, which I received in full from a Shoka [Bhotiya] gentleman who was present as interpreter, is amusing and curious, showing the mutability and hypocrisy of the Tibetans. In the long run, and being well acquainted with the cowardice of his visitors, Mr. Larkin not only obtained redress on every point, but gave the Jong Pen [sic] and his officers a severe harangue. The result of the interview was that the collection of the land revenue should be put a stop to, and that Tibetan law should no more be administered on our side of the frontier" (Landor 1899: 76).

In his view the Tibetans had obviously remained the "masters of the situation", whereas British rule existed only "on paper", especially on maps (Political Department 1896a: 34).[66] The former good feeling of the Bhotiyas toward the colonial government was then depicted as being disrupted "through fear" of the *jongpen's* coercive measures (Political Department 1896e: 24).[67] Worse still, the latter apparently not only claimed fiscal-cum-territorial authority but also judicial authority over the Bhotiyas. It was reported that the traders, whether for convenience or by constraint, did not take their grievances and disputes to the British headquarters in Almora – a 250 km journey – but "to Taklakot, a march beyond our border, where they are in the midst of their own kinsmen at the seat of their trade" (Political Department 1896a: 34). While all these everyday arrangements appeared like grit in the gears of the imperial machinery, and provoked numerous recommendations for improvement, most efforts were directed toward solving the highly controversial issue of land revenue.

All along the *jongpen* articulated his claim to sovereignty in a very straightforward manner: "It is customary with us to realize tax from Darma in the same way as we do from Byans. We will go on doing so: we cannot leave our *kadimi dastur* [ancient custom]" (Political Department 1896g: 25). To Larkin he further emphasized that a British officer, namely O'Beckett, had even personally sanctioned him to do so (Political Department 1896a: 32). When the deputy collector came to inform the *jongpen* of the British government's legal prohibition on the collection of any land revenue by a foreign authority, he met with a recalcitrant opponent who took a lot of persuading to give a written acknowledgement of this new order. Under the threat of arrest the *jongpen* put his signature "on the smallest scrap of paper possible", temporarily refrained from collecting his dues and then "arranged to refer the matter to Lhassa [sic], though he puerilely urged that it meant his recall and disgrace" (Political Department 1896a: 32).

The Bhotiyas were alarmed by all this and expected the Tibetans to respond with hard-hitting measures. Influential traders had already sided with the *jongpen*, claiming that the disputed revenue system was also part of their *kadimi dastur*, or ancient custom, and that its dissolution would make them suffer irrevocable losses (Political Department 1895f: 21–22). In fear of exclusion from the commercial circuit with Tibet the traders even refused to take any advances from representatives of the woolen mills in the plains (Political Department 1897a: 44). In other words, as the struggle for territorial sovereignty gradually gained momentum the forecasts looked

---

[66] A number of British people had made a personal experience of these measures. In 1896, for instance, the Tibetans attacked a British military officer, lieutenant Gaussen, near the Lipu Lekh pass without any consequences (Political Department 1896a: 31; Landor 1899: 75). Landor was even taken captive when he visited western Tibet and only released with the help of the missionaries stationed in Dharchula.

[67] Kishen Singh Rawat, another Pundit explorer, reported that the people in Johar were 'frightened' (Political Department 1896e: 23); the *peshkar* said that the residents of Byans and Darma would 'fear' the orders of the Tibetan authorities (Political Department 1896f: 24). A subordinate of the *jongpen* of Taklakot was even depicted as a 'thorough barbarian' who "punished defaulters with brutal floggings" and "acted with great injustice" (Political Department 1897a: 42).

damaging rather than beneficial from the British perspective. Moreover, chances were high that local residents would attribute any adverse effects to the interference by the colonial government. As a way out of these difficulties, in early summer 1897 the deputy commissioner of Almora proposed two principles for the development of a sound policy, which he called the 'imperial' and the 'mercantile'[68]:

> From the first point the payment of any land tax, whether separate or composite, to a foreign power by British subjects for British land is too objectionable to be allowed to continue. On the other hand, from the second point of view, it becomes a question of importance whether the discontinuance of the dues will not play havoc with our trade with Thibet [sic] and subject our traders to much hardship once they cross the borders. [...] If Thibet were a civilized power, the issues would be much simplified. Unfortunately it is a barbaric power, and the frontier officials with whom we have to treat are uncouth, ignorant, and grasping barbarians. (Political Department 1897a: 44)

Based on the assumption that neither the Bhotiyas nor the Tibetans had any clear idea about the distinction "between a land and a trade revenue", the deputy commissioner issued a new order that was aligned to a more centralized mode of imperial management (Political Department 1897a: 44). This order, which could have enforced the empire's claim to territorial sovereignty, sustained a ban on the collection of any foreign dues within British territory but allowed the Bhotiyas to pay whatever they liked in Tibet itself. Since the Tibetans could still elicit any foregone taxes on their side of the border, the continuation of commerce was considered likely. Even more importantly, it was expected that this policy would prevent the "semi-independent" and "ill-controlled" frontier officials of the Lhasa government from undermining "British prestige" (Political Department 1897a: 45). However, observers anticipated that the new order would, at least initially, require "enforcement at the point of the bayonet" (Political Department 1897a: 45).

Only a few weeks after this order had been issued, the Tibetan authorities closed the Lipu Lekh without further ado and pushed trade via a neighboring pass in far-western Nepal (Political Department 1897b: 46). The *peshkar* reported that neither the *jongpen* nor the Bhotiyas were reconciled with the idea of resolving tax-related activities solely in Tibet (Political Department 1897b: 46). The most obvious reason – although poorly understood by British officials in Kumaon – was the fact that the revenue collection was partly realized as what Polanyi (1957: 262) called an 'administered trade': ritualized transactions through government-controlled channels.[69] Before the Bhotiyas were allowed to cross the high passes they were obliged to exchange local grains and liquor for salt and borax that Tibetan nomads brought into their valleys. Whilst the *jongpen* might have fixed the exchange rates to the traders' disadvantage when compared to the going market prices, the system facilitated the

[68]The early British Empire can be conceived as a mercantile project that was geared toward the expansion of commercial monopolies. After 1858, the acquisition of new territory was legitimized through the imperial objective of civilizing primitive places and peoples.

[69]The ritualization of trade in Kumaon included the ceremonial opening of the annual trading season by Tibetan officials (called *sarji*) and the exchanges of gifts between exclusive trade partners (called *mitra*). Crucial is that these ritualized practices permitted Tibetan authorities to articulate their authority (Brown 1984: 82–89).

overall logistic, economic and political feasibility of cross-border commerce: the risk of transportation was distributed among residents on both sides of the high passes, the overall trade volume was enhanced, and all parties – except for the British – secured a competitive advantage.[70] The Tibetan officials maintained their political-cum-fiscal authority and the traders their longstanding monopoly over the trans-Himalayan goods traffic in Kumaon (Brown 1984: 108–113).[71]

The *jongpen* therefore simply encouraged the Bhotiyas to pay their dues in neighbouring settlements of Nepal or at largely uncontrolled high pastures and encamping grounds in their own valleys. Traders welcomed these opportunities since it allowed them to continue visiting the trade marts on the Tibetan plateau (Political Department 1897e: 1). Moreover, Tibetan officials informed the *peshkar* that those traders who did not pay their dues would later have to pay two or three times as much (Political Department 1897f: 2). In other words, the *jongpen* did not "consider the least about concession and civility for the British Government", but instead thought "himself very powerful" (Political Department 1897f: 2). And, indeed, he retained his share. The lieutenant governor of the North-western Provinces rejected the use of military force and instead put forward another proposal with which both the Tibetan authorities and the Bhotiyas were happy to comply. This provided for the maintenance of the old Tibetan system, the collection of dues on British territory, with the exception of being officially redefined as not including land revenue and as supervised by the *political peshkar* (Political Department 1897g: 7).[72] Subsequently trade not only resumed but also flourished, as demonstrated by a near doubling of imported wool between 1897 and 1900 (see Table 2.1) (Goudge 1901: 3). While the outcome was thus clearly successful for the colonial economy, the question remains as to its implications in terms of sovereignty and the application of imperial law.

The issue of confluent territories was reconciled, at least from a British point of view, by enforcing a strict separation between land taxes and trade duties. However, this solution also required the British to accommodate themselves to overlapping sovereignties. On the one hand they officially permitted the Tibetan *jongpen* to continue exerting his influence on the Bhotiyas' commerce by collecting trade dues on British territory. On the other hand, and as a preventive measure to any unauthorized Tibetan claim over the empire's territory, the colonial government also strengthened the power of other political actors. At the outset of the dispute British military men had proposed a reinforcement of the empire's sovereign claim over the Kumaon Himalaya through the posting of "a strong official from some other district" in the

[70] The *jongpen* argued that the proposed change would have greatly reduced "the supplies (of food) and liquor he got up to present" (Political Department 1897c: 50).

[71] While British officials in Kumaon depicted the traditional modes of exchange as "forced trade", the chief secretary of the North-western Provinces regarded it as a locally modified version of "free exchange" (Political Department 1897d: 53).

[72] Administrators in Kumaon had initially ordered the *peshkar* to arrest any Tibetan official who would try to collect land revenue from the Bhotiyas. However, without any external support Kharak Singh did not dare to put this order into practice, not least since local residents often took a stand for the Tibetan trade agents (Political Department 1897h: 47).

Bhotiya valleys (Political Department 1894a: 26).[73] That Kharak Singh Pal, a representative of Kumaon's powerful aristocracy, was chosen for this duty and permanently placed on the empire's payroll was perhaps a far more effective strategic move (Political Department 1897i: 7). With the simultaneous nomination of the Rajwar of Askot as a native judge, the British ensured for the first time since their arrival in Kumaon that both the executive and judicial branches of the empire's civil service were permanently present and within easy reach of the Bhotiyas. As well respected local figures these two men were in a position to gradually enforce imperial law and thus alter the Bhotiyas' cool indifference to British officialdom. An initial step in this direction was accomplished by granting these new imperial officials the right to impose and collect a 'rent' of 4 *annas* for each house inhabited by the Bhotiyas during the winter season (Political Department 1895c: 32). In other words, whilst the Bhotiyas still maintained their trade monopoly they had lost their privileged status as political 'ambassadors'. The traders' privileges were superseded by the ruling family of Askot, whose primary task was to move a rather patchy jurisdictional net towards a more uniform and formal rule of law.

2.7 Coda

Throughout the nineteenth century the Bhotiyas' trade was characterized by growth. This situation was set to change in the twentieth century. The trade in borax (*tincal*), a mineral from the saline lakes of Central Tibet that was utilized by European porcelain industries, already declined during the second half of the nineteenth century. Reasons included the discovery of new deposits in other parts of the world as well as the development of chemical processes for producing synthetic substitutes (Brown 1984: 117–119; van Spengen 2000: 101). From the 1920s onwards, Tibetan wool was successively replaced by cheaper imports from Europe and Australia (Roy 2003: 258–259). In addition, the trade in non-borate salts became less important due to the wide availability of sea salt from the coastal areas of India (von Fürer-Haimendorf 1975: 62). In the course of these developments the main commercial routes to Tibet shifted eastwards, linking Sikkim with the ports of Bengal (Foreign Department 1947), thus precluding the Bhotiyas' access to the main trading networks. The situation further worsened for them because of a widespread outbreak of animal diseases among their flocks during the 1940s. Intensifying political disputes – India struggling for independence from the British Raj and Tibet defending its autonomy vis-à-vis the Republic of China – seemed to finally herald a

[73] This proposal was justified by the observation "that the Bhootias [sic] themselves are very high-handed with the villagers on their way down country" (Political Department 1894a: 26). However, the selection of new staff was confined by the remoteness of the place, which most outsiders were not willing to accept voluntarily, and the language competences that were required to understand and efficiently handle local affairs (Political Department 1895b: 26–27).

halt to the process of imperial transformation and, with it, the forms of cross-border commerce that it had boosted and constantly kept in motion.

After India's independence in 1947 and after the Communist Party had taken power in China in 1949, a period of cautious cooperation between these newly founded nation states began. Anti-colonial and anti-imperialist movements as well as the idea of a prosperous and self-governed Asia assured a basis of confidence (Guha 2007: chapter 8). In 1951 the Government of India recognized China's control in Tibet, allowing both countries to formalize their friendship in a contract (Five Principles of Peaceful Coexistence – *Panch Shila*) and to sign a trade agreement regulating commerce and pilgrimage traffic across the shared border in 1954. Disagreements were smoothed over at that time (GoI 1960), not least because Indian state officials conceived the High Himalaya as an effective natural barrier that would prevent any serious threat to national concerns (Nehru 1985 [1952]: 74). This position was mirrored in the minimal expenditures that India spent on developing this mountain region (Gerwin and Bergmann 2012).

The Chinese government, in contrast, initiated a large infrastructural program immediately after their invasion of Tibet in 1950 (Ginsburgs and Mathos 1960). Politically the most critical intervention of this kind was the construction of a road through the Aksai Chin area of Ladakh, which was only discovered by the Indian administration towards the end of the 1950s (Kreutzmann 2007). This fueled an increasingly adversarial situation on the border. After several inconclusive rounds of negotiation between the two Prime Ministers, Jawaharlal Nehru and Chou En Lai, a joint Sino-Indian committee was established in 1960 (Kirk 1962: 139). Officials and advisors from both governments were to

> examine, check and study all historical documents, records, accounts, maps and other material relevant to the boundary question, on which each side relied in support of its stand and draw up a report for submission to the two Governments. (GoI 1961: 1)

The commission conducted a sector-wise discussion of the shared boundary.[74] Kumaon and Garhwal fell into the 'Middle Sector', where its proposed course differed only slightly in three areas inhabited by the Bhotiyas, namely Nilang-Jadhang and Niti-Barahoti in Garhwal as well as Sangchamalla-Lapthal in Kumaon. Both parties agreed on the general existence of what was called a 'traditional customary boundary line' (TCBL), but held different views on what they deemed valid as 'factual' evidence for determining and making explicit its 'true' course. The Indian side based its overall position on what they perceived as an 'innate' and 'self-evident' feature, viz. the watershed-line, and argued that the TCBL

> takes shape on the basis of the natural features of the area, and is later recognised through a process of acknowledgement spread over centuries of custom and tradition. Much later, it may be confirmed by delineation, demarcation, definition or even formal delimitation, but [...] such confirmation is not necessary to its validity. Formal agreements, though essential for artificial boundaries, are optional in the case of a boundary based on natural features,

---

[74] Namely (a) the western sector with Kashmir; (b) the middle sector that included the Punjab, Himachal Pradesh, and Uttarakhand (then part of Uttar Pradesh); and (c) the eastern sector of the North-East Frontier Agency (NEFA).

> which had been traditionally recognised. [...] Such boundaries do not naturally change and if they change, they become artificial boundaries. Certainly the strength and weakness of the States concerned or the exercise of effective authority or military control in the border areas do not in themselves affect legitimate title or result in any change in the location of a traditional line. (GoI 1961: 286)[75]

The Chinese discredited the whole line of argument, especially by disclosing many inconsistencies between the deployed principle and the alignments that were actually claimed by India. In consequence they picked on the determinist outlook and general inadequacy of the watershed-principle and instead emphasized that the TCBL

> was formed gradually in a long process of historical development according to the extent up to which each side all along exercised administrative jurisdiction, and therefore it is inconceivable that in the early periods of history, this line was already fixed as the present alignment. [...] As for the question of precision, as the Sino-Indian traditional customary line has not been formally delimited and the two governments never conducted any joint surveys, it cannot be very precise at every point, but can only be an approximate line. Of course, if one were to provide every detail along the alignment it advocates merely according to the map, that would not be impossible for either side, but what would be the meaning of this? Can it be ensured that information provided in such a fashion is definitely in conformity with the actual situation on the ground? (GoI 1961: CR-178)

In other words, Chinese diplomats opposed the idea of a boundary based on natural features by foregrounding the historical developments that shaped specific territorial arrangements but still awaited formal delimitation. Both India and China were keen to substantiate their claims through older cartographic visualizations, a procedure that the Indian side tellingly depicted as a 'war of maps' (GoI 1961: 233–259; cf. Kirk 1962). However, the dispute did not stop there! War broke out in October 1962, and by November the Chinese army had broken down almost all Indian defenses, after which they suddenly and unilaterally declared a cease-fire. They moved back behind the MacMahon-Line in the northeast, but continued to occupy Aksai Chin. The resulting new boundary later became officially known as the Line of Actual Control (LAC). Soon after the war, Indian officials began intensive discussions about military strategies and failures, which were picked up by the media and regional politicians. Subsequently the military budget was drastically enhanced, making it possible to secure the Himalayan border region that had then become overshadowed not only by the conflict with China but also with Pakistan. It was during this time that the Border Road Organization (BRO), the Indo-Tibetan Border Police (ITBP) and secret agencies surveying these areas were founded.

One of the most consequential interventions by the Indian state in the Himalayan region was an extensive road-building program. This program, which was promoted by Indian militaries, opened the lower parts of the valleys inhabited by the Bhotiyas

---

[75] In the course of the Himalayan border conflict, the Indian side took custom and tradition as a passive receptor for what would have been "defined by nature, confirmed by history and sanctified by the laws of nations" (GoI 1961: 287) since times immemorial, viz. the watershed-line. For that reason even the ancient Indian epics were considered as depicting the TCBL in its present form (MEAI 1959: 125–132).

to heavy vehicles. Most roads were built immediately in the years following the war, and their total length in Kumaon "increased from 360 km in 1947 to 6,421 km in 1991" (Rawat and Sharma 1997: 118). This renewed government effort to control the region also became noticeable in a rise of rural development programs. School building activities, as well as the implementation of an enlarged administrative structure in the mountain area of Kumaon indicate this trend. In the Bhotiya valleys most of the existing government and public facilities, such as Block Development Offices, hospitals or secondary schools, were established in the years of drastically enhanced state presence following the border war.

Van Schendel (2007: 37) argues that the explosive nature and continuing relevance of the Sino-Indian boundary dispute stems largely "from failed border-making exercises in the colonial period". I would suggest further that the conflict was also nourished by a rather persistent presence of confluent territories and overlapping sovereignties. The associated negotiations have been examined on the basis of an analytical approach that is anchored in van Schendel's proposal to conceive of High Asia as a continuous zone and an agentive site of political action, which he named Zomia (van Schendel 2002). Whereas Scott further conceptualized Zomia as a 'non-state space' with regard to the mountain regions of Southeast Asia, I have followed Shneiderman's suggestion to envisage the Himalaya as a 'multiple-state space' (Shneiderman 2013). Shneiderman adopts this kind of 'Zomia-thinking' to analyze processes of state- and political subject-formation in the contact zones between polities that remained independent throughout the colonial era, namely Nepal and Tibet. Although the histories and geographies of European colonization and frontier imperialism also played a role in these settings, the case of Kumaon provides the alternative analytical focus of a Non-regulation Province within Britain's nineteenth-century Indian Empire. This focus has allowed me to illustrate that confluent territories and overlapping sovereignties are key to understanding imperial frontiers. Whilst the empire's higher-level elites often imagined their sovereign claims as being inherently territorial, articulations of imperial power and authority within the considered localized setting were marked by a great uncertainty as to both who was in control of what and where. In other words, British imperial sovereignty, similar to that of older regimes in the area, remained malleable and contested within the wider relational field of the Bhotiyas' trans-Himalayan trade.

The findings of this chapter clearly show the limits of British imperial sovereignty in the Himalaya. During the 1840s the commissioner accepted these limits owing to the priority of securing access to the Tibetan plateau and its commercial goods through the Bhotiyas' loyalty. Initially these limits were recognized primarily in terms of the strict prohibition on the entry of British officials into Tibet. In the 1860s the British undermined this closed-door policy by dispatching specially trained Bhotiya recruits to survey the Tibetan plateau for them. The resulting maps played a key role as to how British officials imagined their political authority, namely as being defined territorially. However, in the 1890s it gradually transpired that there still existed a great uncertainty as to who was in control over what and where, which was by then no longer considered acceptable. The continued existence of such limits to imperial sovereignty was evidenced through the collection of land taxes on British territory by the Tibetan authorities, which the Bhotiya traders partly supported for their own strategic interest. The British solved the resulting problem of confluent territories by enforcing a strict separation between land taxes and trade duties, thereby accommodating imperial power and authority to overlapping sovereignties. Whilst the Tibetans were officially allowed to continue the collection of trade dues from the Bhotiyas on British territory, local aristocrats were promoted in order, firstly, to prevent any unauthorized Tibetan claim to power over the empire's territory and, secondly, to alter the Bhotiyas' cool indifference toward the colonial government.

The study, therefore, provides some qualification to Scott's analysis of Zomian traders (Scott 2009: 4–9). Whilst Scott generally foregrounds their agency in retaining local autonomy by intentionally resisting projects of state and empire building, the picture is different in the specific case of the Bhotiyas. These trans-Himalayan traders were adept not only at staying close to Britain's nineteenth-century Indian Empire, but also in promoting it as a means to raise their own profits. Moreover, it was the British who nurtured the Bhotiyas' relative autonomy in order to gain their loyalty and therewith ensure that imperial interests were satisfied. In other words, the study highlights both the agentive participation of trans-Himalayan traders in the imperial project and the empire's necessary flexibility in articulating sovereignty at its high altitude frontier.

Moreover, whilst Scott hesitates to pursue his analysis of Zomia into the postcolonial present, I would rather argue with Ludden (2011: 133; cf. 2012) that it is important to understand how "imperial frontiers extend across boundaries of time into present-day national territory". The reproduction of imperial forms, which typically revolves around questions of location, gender, ethnicity and class became adjusted to and persisted in emerging geographical frames, such as the nation state. Such a scalar reframing of imperial priorities not only relates to a state's approach in building a nation and bringing regions and peoples under its wing. They also concern the ways in which local groups, such as the Bhotiyas, make use of their longstanding and intimate acquaintance with the forces through which the Himalayan border region evolved as an imperial environment. How have they utilized this acquaintance to assert who they are or want to be within the emergent configurations of postcolonial India? It is such questions of identity that will be dealt with in the following chapter.

References

Agnew J (2005) Sovereignty regimes: territoriality and state authority in contemporary world politics. Ann Assoc Am Geogr 95:437–461

Alder GJ (1980) Standing alone: William Moorcroft plays the great game, 1808–1825. Int Hist Rev 2(2):172–215

Alder GJ (1985) Beyond Bokhara. The life of William Moorcroft, Asian explorer and pioneer veterinary surgeon. Century Publishing, London, pp 1767–1825

Allen J (2011) Topological twists: power's shifting geographies. Dialogues Hum Geogr 1(3):283–298

Anand D (2009) Strategic hypocrisy: the British imperial scripting of Tibet's geopolitical identity. J Asian Stud 68(1):227–252

Anon (1817) Himalaya mountains and Lake Manasawara. Q Rev 17:403–441

Assmann J (2011) Cultural memory and early civilization: writing, remembrance, and political imagination. Cambridge University Press, Cambridge

Barrow IJ (2003) Making history, drawing territory. British mapping in India, c. 1756–1905. Oxford University Press, New Delhi

Bateson G (1935) Cultural contact and schismogenesis. Man 35:178–183

Bateson G (1958) Naven, 2nd edn. Stanford University Press, Stanford

Batten JH (1851) Final report on the settlement of Kumaon, 1848. In: Batten JH (ed) Official reports on the province of Kumaon with a medical report on the Mahamurree in Gurhwal, in 1849–1850. Secundra Orphan Press, Agra, pp 258–356

Bayly CA (1996) Empire and information: intelligence gathering and social communication in India, 1780–1870. Cambridge University Press, Cambridge

Benton L (2008) From international law to imperial constitutions: the problem of quasi-sovereignty, 1870–1900. Law Hist Rev 26(3):595–619

Benton L (2010) A search for sovereignty: law and geography in European empires 1400–1900. Cambridge University Press, Cambridge

Bergmann C (2016) Confluent territories and overlapping sovereignties: Britain's nineteenth-century Indian empire in the Kumaon Himalaya. J Hist Geogr 51:88–98

Bergmann C, Gerwin M, Sax WS, Nüsser M (2011) Politics of scale in a high mountain border region: being mobile among the Bhotiyas of the Kumaon Himalaya, India. Nomadic Peoples 15(2):104–129

Berreman GD (1963) Hindus of the Himalayas: ethnography and change. University of California Press, Berkeley

Bishop P (1989) The myth of Shangri-la: Tibet, travel writing and the western creation of sacred landscape. Athlone Press, London

Blackburn S (2003) Colonial contact in the 'hidden land': oral history among the Apatanis of Arunchal Pradesh. Indian Econ Soc Hist Rev 40(3):335–365

Bray J (2009) Krishnakanta Basu, Rammohan Ray and early 19th century British contacts with Bhutan and Tibet. Tibet J 34(3)–35(2):329–356

Brown CW (1984) The goat is mine, the load is yours: morphogenesis of 'Bhotiya-Shauka', U.P., India. Lund University, Department of Social Anthropology, Lund

Brown CW (1987) Ecology, trade and former Bhotiya identity. In: Raha MK (ed) The Himalayan heritage. Gian Publishing House, New Delhi, pp 125–138

Browne ECM (n.d.) Among the Bhotiyas and their neighbors. Nichols Press, Lynn

Burghart R (1992) Gifts to the gods: power, property and ceremonial in Nepal. In: Cannadine D, Price S (eds) Rituals of royalty: power and ceremonial in traditional societies. Cambridge University Press, Cambridge, pp 237–270

Charak SS (2003 [1983]) Extracts from 'General Zorawar Singh'. In: McKay A (ed) The history of Tibet, vol II. The medieval period, c. 850–1895, the development of Buddhist paramountcy. Routledge Curzon, London, pp 748–766

Clarke J (1998) Hindu trading pilgrims. In: McKay A (ed) Pilgrimage in Tibet. Curzon, Richmond, pp 52–70

Cohen B (1985) The command of language and the language of command. In: Guha R (ed) Subaltern studies IV. Oxford University Press, New Delhi, pp 276–329

Cohn BS (1964) The role of the gosains in the economy of eighteenth and nineteenth century upper India. Indian Econ Soc Hist Rev 1:175–182

Connerton P (1989) How societies remember. Cambridge University Press, Cambridge

Dangwal DD (2009) The lost mobility: pastoralism and modernity in Uttarakhand Himalaya (India). Nomadic Peoples 13(2):84–101

Datta CL (1969 [1968]) Zorawar Singh: the political mission of J.D. Cunningham, 1841–1842. Bengal Past Present 88:82–89
de Laet M, Mol A (2000) The Zimbabwe bush pump: mechanics of a fluid technology. Soc Stud Sci 30:225–263
Dhakriyal DS (2004) Himalayi Sauka sanskritik dharohar, vol I. Takshila Prakashan, New Delhi
Dirks N (1993) Colonial histories and native informants: biography of an archive. In: Breckenridge CA, van der Veer P (eds) Orientalism and the postcolonial predicament (44th annual South Asia seminar at the University of Pennsylvania, 1988/1989). University of Pennsylvania Press, Philadelphia, pp 279–313
Edney MH (1997) Mapping an empire. The geographical construction of British India. University of Chicago Press, Chicago, pp 1765–1843
Fisher JF (1978) Introduction. In: Fisher JF (ed) Himalayan anthropology: the Indo-Tibetan interface. Mouton, The Hague, pp 1–3
Fisher JF (1986) Trans-Himalayan traders: economy, society, and culture in northwest Nepal. University of California Press, Berkeley
Foreign Department (1947) Annual report of the British trade agent, Gartok. PS/12/4209 Wool Trade 25th Feb. 1942–30th June 1947. India Office Library and Records at the Asia and Africa Collection of the British Library, London [hereafter IOL]
Forest Department (1905) Petition received from certain Bhotias complaining of the insufficiency of grazing near Ramnagar, 25 April 1905. Commissioner Office Records, 61/1903–1907, Box 147. Regional Archive Uttarakhand, Nainital, India [hereafter RAU]
Francke AH (1907) A history of western Tibet. One of the unknown empires. Partridge, London
Gerwin M, Bergmann C (2012) Geopolitical relations and regional restructuring: the case of the Kumaon Himalaya, India. Erdkunde 66(2):1–17
Ginsburgs G, Mathos M (1960) Communist China's impact on Tibet. The first decade I. Far East Surv 29(7):102–109
Goffman E (1959) The presentation of self in everyday life. Anchor Books, New York
GoI [Government of India] (1960) Text of the letter of February 5, 1960, from prime minister of India to the prime minister of China and text of the note of the government of India of February 12, 1960. Ministry of External Affairs, New Delhi
GoI [Government of India] (1961) Report of the officials of the governments of India and the People's Republic of China on the boundary question. Ministry of External Affairs, New Delhi
Goldstein M (1989) A history of modern Tibet, 1913–1951: the demise of the lamaist state. University of California Press, Berkeley
Goudge JE (1901) Report on the assessment of pargana Darma. RAU
Goudge JE (1903) Almora: final report on the assessment of the Almora district and the hill pattis of the Naini Tal district. V/27/314/643, IOL
Guha R (1989a) The unquiet woods. Ecological change and peasant resistance in the Himalaya. Oxford University Press, New Delhi
Guha R (1989b) Dominance without hegemony and its historiography. In: Guha R (ed) Subaltern studies VI. Writings on South Asian history and society. Oxford University Press, New Delhi, pp 210–309
Guha R (1999) Environment and ethnicity in India, 1200–1991. Cambridge University Press, Cambridge
Guha R (2001) The prehistory of community forestry in India. Environ Hist 6(2):213–238
Guha R (2007) India after Gandhi. The history of the world's largest democracy. Ecco, New York
Harris T (2013) Geographical diversions: Tibetan trade, global transactions. University of Georgia Press, Athens
Hastings SF (ed) (1858) The private journal of the marquess of Hastings, governor-general and commander in chief in India, 2 vols. N.p., London
Hitchcock J (1966) The Magars of Banyan hill. Holt Rinehart and Winston, New York
Holdich T (1899) The use of practical geography illustrated by recent frontier operations. Geogr J 13(5):465–477

Hopkirk P (1982) Trespassers on the roof of the world. The race for Lhasa. John Murray, London
Huttenback RA (1961) Gulab Singh and the creation of the Dogra state of Jammu, Kashmir and Ladakh. J Asian Stud 20(4):477–488
Huttenback RA (1968) Kashmir as an imperial factor during the reign of Gulab Singh (1846–1857). J Asian Hist 2:77–108
Inden R (1981) Hierarchies of kings in early medieval India. Contrib Indian Sociol 15(1–2):99–125
Inden R (2000) Imagining India. Indiana University Press, Bloomington
EIC [East India Company] (1983 [1824]) Papers respecting the Nepal war. Bimla Publishing House, New Delhi
Irwin J (1973) The Kashmir shawl. Her Majesty's Stationery Office, London
Jackson D (1976) The early history of Lo (Mustang) and Ngari. Contrib Nepal Stud 4(1):39–56
Jackson D (1978) Notes on the history of Se-rib and nearby places in the upper Kali Gandaki. Kailash 6:195–228
Joshi MP, Brown CW (1987) Some dynamics of Indo-Tibetan trade through Uttarakhand (Kumaon-Garhwal), India. J Econ Soc Hist Orient 30(3):303–317
Kak M (2004) Indo-Tibet trade routes as public sphere. In: Muthukumaraswamy MD, Kaushal M (eds) Folklore, public sphere and civil society. Indira Gandhi National Centre for the Arts, New Delhi, pp 138–156
Kipling R (n.d.) Kim. Alex Catalogue, eBook Collection, EBSCOhost. Accessed 18 Aug 2015
Kirk W (1962) The inner Asian frontier of India. Trans Pap (with the Inst Br Geogr) 31:131–168
Kreutzmann H (2007) Grenzen und Handel. Geopolitische Implikationen im indisch-chinesischen Spannungsfeld. Geograpahische Rdsch 59(1):4–11
Kumaon Division (1842a) G.T. Lushington to the lieutenant governor N.W.P., 26 March, Political letters issued, vol 50. State Archive Uttar Pradesh, Lucknow, India [hereafter SAUP].
Kumaon Division (1842b) G.T. Lushington to R.N.C. Hamilton, 16 August, Political letters issued, vol 50. SAUP
Kumaon Division (1842c) Translation of a report from the putwary of Bians, 28 August, forwarded 14 September from G.T. Lushington to lieutenant governor N.W.P., Political letters issued, vol 50. SAUP
Kumaon Division (1842d) G.T. Lushington to R.N.C. Hamilton, 14 September, Political letters issued, vol 50. SAUP
Kumaon Division (1842e) G.T. Lushington to R.N.C. Hamilton, 17 September, Political letters issued, vol 50. SAUP
Kumaon Division (1842f) G.T. Lushington to the secretary of government N.W.P., Secret Department, 9 February, Political letters issued, vol 50. SAUP
Kumaon Division (1842g) G.T. Lushington to R.N.C. Hamilton, 27 April, Political letters issued, vol 50. SAUP
Kumaon Division (1842h) Translation of a report from putwary of Bians, 11 September, forwarded 23 September by G.T. Lushington to R.N.C. Hamilton, Political letters issued, vol 50. SAUP
Kumaon Division (1842i) G.T. Lushington to R.N.C. Hamilton, 24 October, Political letters issued, vol 50. SAUP
Kumaon Division (1843a) G.T. Lushington to H.M. Elliot, Esq., Revenue letters issued, vol 15. SAUP
Kumaon Division (1843b) G.T. Lushington to R.N.C. Hamilton, 30 August, Political letters issued, vol 50. SAUP
Lamb A (1958) Tibet in Anglo-Chinese relations: 1767–1842. J R Asiat Soc GB Irel 1–2:26–43
Lamb A (1986) British India and Tibet: 1766–1910. Routledge, London
Lamb A (ed) (2002) Bhutan and Tibet: the travels of George Bogle and Alexander Hamilton 1774–1777, vol 1., letters, journals and memoranda. Roxford Books, Hertingfordbury
Landor AH (1899) In the forbidden land: an account of a journey in Tibet, capture by the Tibetan authorities, imprisonment, torture, and ultimate release. William Heinemann, London

Law J (1986) On the methods of long distance control: vessels, navigation, and the Portuguese route to India. In: Law J (ed) Power, action and belief: a new sociology of knowledge? [Sociological Review Monograph, vol 32]. Routledge, London, pp 234–263
Law J, Mol A (2001) Situating technoscience: an inquiry into spatialities. Environ Plan D Soc Space 19(5):609–621
Leach E (1961) Rethinking anthropology. Athlone Press, London
Leder S (2003) Zu einer Zeit, als Bäume und Gräser noch sprechen konnten: Sozioökonomie, Kosmologie und Mythologie der Rang-Shauka im zentralindischen Himalaya. Peter Lang, Frankfurt am Main
Legg S (2009) Of scales, networks and assemblages: the league of nations apparatus and the scalar sovereignty of the Government of India. Trans Inst Br Geogr 34:234–253
Legg S (2014) An international anomaly? Sovereignty, the league of nations, and India's princely geographies. J Hist Geogr 43:96–110
Lester A (2012) Humanism, race and the colonial frontier. Trans Inst Br Geogr 37(1):132–148
Levi-Strauss C (1985) The jealous potter. University of Chicago Press, Chicago
Lewis TT (1994) Himalayan regions in comparative perspective: considerations regarding Buddhism and Hinduism across their Indic frontier. Himal Res Bull 14(1–2):25–46
Ludden D (2011) The process of empire: frontiers and borderlands. In: Bang PF, Bayly CA (eds) Tributary empires in global history. Palgrave Macmillan, London, pp 132–150
Ludden D (2012) Imperial modernity: history and global inequity in rising Asia. Third World Q 33(4):581–601
MacGregor J (1970) Tibet: a chronicle of exploration. Routledge, London
Mantena K (2010) Alibis of empire: Henry Maine and the ends of liberal imperialism. Princeton University Press, Princeton
Manzardo AE (1978) To be kings of the north: community, adaptation, and impression management in the Thakali of western Nepal. University of Wisconsin-Madison, Madison, PhD-dissertation
Marczell PJ (1997) William Moorcroft's commercial pilgrimage to Manasa Sarowara in 1812. In: Osmaston H, Tsering N (eds) Recent research on Ladakh, vol 6. Proceedings of the sixth international colloquium on Ladakh, Leh, 1993. Motilal Banarsidass Publishers, New Delhi
Markham CR (1875) Travels in great Tibet, and trade between Tibet and Bengal. J R Geogr Soc Lond 45:299–315
Markham CR (1876) Narratives of the mission of George Bogle to Tibet, and of the journey of Thomas Manning to Lhasa [edited with notes, an introduction, and lives of Mr. Bogle and Mr. Manning]. Trübner, London
Markham CR (1968 [1878]) A memoir on the Indian surveys. Meridian Publishing Co., Amsterdam
Mason K (1923) Kishen Singh and the Indian explorers. Geogr J 62(6):429–440
McConnell F (2013) Sovereignty. In: Dodds K, Kuus M, Sharp J (eds) The Ashgate research companion to critical geopolitics. Ashgate, Farnham, pp 109–128
McGranaham C (2003) From Shimla to Rongbatsa: the British and the 'modern' boundaries of Tibet. Tibet J 28(4):39–60
McKay A (1992) The establishment of the British trade agencies in Tibet: a survey. J R Asiat Soc G B Irel (Third Series) 2(3):399–421
McKay A (1997) Tibet and the British raj: the frontier cadre, 1904–1947. Curzon, Richmond
McKay A (2003) 19th century British expansion on the Indo-Tibetan frontier: a forward perspective. Tibet J 28(4):61–76
MEAI [Ministry of External Affairs] (1959) White paper no. II. Notes, memoranda and letters exchanged between the governments of India and China, September–November 1959, and a note on the historical background of the Himalayan frontier of India. N.p., New Delhi
Metcalf TR (2007) Imperial connections. India in the Indian ocean arena, 1860–1920. University of California Press, Berkely
Michaels A (2001) Traditional Sanskrit learning in contemporary India. In: Michaels A (ed) The pandit: traditional scholarship in India. Manohar, New Delhi, pp 3–16

Misra BB (1977) The bureaucracy in India. An historical analysis of development up to 1947. Oxford University Press, New Delhi

Mittal AK (1986) British administration in Kumaon Himalayas. Mittal Publications, New Delhi

Mol A, Law J (1994) Regions, networks and fluids: anaemia and social topology. Soc Stud Sci 24:641–671

Montgomerie TG (1868) Report of a route survey made by pundit, from Nepal to Lhasa, and thence through the upper valley of the Brahmaputra to its source. J R Geogr Soc 38:129–210

Montgomerie TG (1869) Report of the trans-Himalayan explorations during 1867. Proc R Geogr Soc Lond 13(3):1868–1869

Moorcroft W (1818) A journey to Lake Manasarovar in Undes, a province of little Tibet. Asiat Res 12:380–536

Moorcroft W, Trebeck G (1976 [1837]) Travels in the Himalayan provinces of Hindustan and the Panjab, vol 1. Faran Academy, Lahore

Murdoch J (2006) Post-structuralist geography: a guide to relational space. Sage, London

Nehru J (1985 [1952]) Letter to the chief minister, dated 2 August, 1952. In: Parthasaranti G (ed) Jawaharlal Nehru – letters to chief minister 1947–1967, vol 3, 1952–1954. Jawaharlal Nehru Memorial Fund, New Delhi

Nüsser M (2006) Ressourcennutzung und nachhaltige Entwicklung im Kumaon-Himalaya (Indien). Geogr Rundsch 58(10):14–22

Obeyesekere G (1992) The apotheosis of captain Cook: European mythmaking in the Pacific. Princeton University Press, Princeton

Ogborn M (2007) Indian ink: script and print in the making of the English East India Company. University of Chicago Press, Chicago

Ogborn M (2012) Review: Lauren Benton. A search for sovereignty: law and geography in European empires, 1400–1900. Am Hist Rev 117:814–816

Onley J (2009) The raj reconsidered: British India's informal empire and spheres of influence in Asia and Africa. Asian Affairs 40(1):44–62

Pande G, Bergmann C (2009) Dharchula: atit ke aine me. Pahar 16:317–324

Pearse H (1905) Moorcroft and Hearsey's visit to Lake Mansarowar in 1812. Geogr J 26(2):180–187

Pemble J (2008) Britain's Gurkha war. The invasion of Nepal, 1814–16. Frontline Books, London

Pignède B (1993 [1966]) The Gurungs: a Himalayan population of Nepal. Ratna Pustak Bhandar, Kathmandu

Piper K (2002) Cartographic fictions: maps, race and identity. Rutgers University Press, New Brunswick

Polanyi K (1957) The economy as instituted process. In: Polanyi K, Arensberg CM, Pearson HW (eds) Trade and early market in the early empires: economies in history and theory. Henry Regnery Company, Chicago, pp 243–269

Political Department (1894a) Lieutenant C.E. Bliss to the deputy assistant adjutant-general Rohilkhund district, 16 Nov., N.W.P. and Oudh, Proceedings for January 1895 (No. 10). RAU, pp 25–27

Political Department (1894b) Major-general Channer, C.B., V.C., commanding Rohilkhund district, to the deputy commissioner of Almora, 6 Oct., N.W.P. and Oudh, Proceedings for October 1894 (No. 14). RAU, pp 5–6

Political Department (1895a) Rajwar of Askot, Almora District, Commissioner Office Records, 24/1896–1867. RAU

Political Department (1895b) Deputy commissioner of Almora to commissioner, Kumaon division, n.d., N.W.P. and Oudh, Proceedings for October 1895 (No. 16). RAU, pp 23–29

Political Department (1895c) Letter to the commissioner, Kumaon division, communicating order on serial no. 40, and sanctioning the entertainment of an establishment costing Rs. 942 a year for one year from 1st October 1895, 11 October, N.W.P. and Oudh, Proceedings for October 1895 (No. 17). RAU, pp 31–33

Political Department (1895d) From the commissioner, Kumaon division, submitting the deputy commissioner of Almora's report on the result of his visit to Garbyang, along with proposals for placing the administration of the outlying tracts in that district on a proper footing, 12 August, N.W.P. and Oudh, Proceedings for October 1895 (No. 15). RAU, pp 21–23
Political Department (1895e) Deputy commissioner of Almora to commissioner, Kumaon division, 31 January, N.W.P. and Oudh, Proceedings for April 1895 (No. 8). RAU, pp 4–5
Political Department (1895f) Translation of Pashkar Karakh Singh's report, 26 December, N.W.P. and Oudh, Proceedings for August 1897 [No. 26[c]). RAU, pp 21–22
Political Department (1896a) Deputy collector of Almora to deputy commissioner of Almora, Mr. Larkin's report on the subject of the dues levied by the Thibetans from British subjects, 15 October, N.W.P. and Oudh, Proceedings for August 1897 (No. 28). RAU, pp 26–37
Political Department (1896b) From the commissioner, Kumaun division, submitting certain reports received from the deputy commissioner of Almora on the subject of dues realized by the Thibetans from British subjects in pattis Byans, Chaudans, and Malla Darma, 12 November, N.W.P. and Oudh, Proceedings for August 1897 (No. 26[a]). RAU, pp 19–20
Political Department (1896c) Translation of a report from Peshkar Dalip Singh of Almora Tahsil to the deputy commissioner of Almora, 16 July, N.W.P. and Oudh, Proceedings for August 1897 (No. 26[f]). RAU, p 23
Political Department (1896d) Deputy commissioner of Almora to commissioner, Kumaon division, 29 October, N.W.P. and Oudh, Proceedings for August 1897 (No. 27). RAU, pp 25–26
Political Department (1896e) Translation of a report from Rai Kishan Singh Rawat, Bahadur, to deputy commissioner of Almora, 25 June, N.W.P. and Oudh, Proceedings for August 1897 (No. 26[d]). RAU, pp 22–23
Political Department (1896f) Translation of Kharak Singh's report, 28 July, N.W.P. and Oudh, Proceedings for August 1897 (No. 26[h]). RAU, p 24
Political Department (1896g) Translation of jongpen's statement, 19 August, N.W.P. and Oudh, Proceedings for August 1897 (No. 26[m]). RAU, p 25
Political Department (1897a) Deputy commissioner of Almora to commissioner, Kumaon division, 25 June, N.W.P. and Oudh, Proceedings for August 1897 (No. 32). RAU, pp 41–46
Political Department (1897b) Translation of a report from peshkar Kharak Singh Pal, 29 June, N.W.P. and Oudh, Proceedings for August 1897 (No. 34). RAU, p 46
Political Department (1897c) Translation of a report from peshkar Karakh Singh Pal to deputy commissioner of Almora, 5 August, N.W.P. and Oudh, Proceedings for August 1897 (No. 48). RAU, p 50
Political Department (1897d) To commissioner, Kumaon division, stating, in reply to serial no. 68, that a distinction should be made between dealings by way of barter and the collection of fixed dues claimed as revenue by the Thibetans, and that nothing should be done to prevent a free exchange of grain and salt or other articles of trade by way of barter, 31 August, N.W.P. and Oudh, Proceedings for August 1897 (No. 51). RAU, p 53
Political Department (1897e) Government of India, foreign department, communicating remarks on the subject of dues levied or claimed by Thibetans from the Bhutias of Byans and Chaudans in Almora district, 18 September, N.W.P. and Oudh, Proceedings for December 1897 (No. 1). RAU, p 1
Political Department (1897f) Translation of peshkar Kharak Singh's report, 19 Aug., with deputy commissioner's orders thereon, 27 August, and the peshkar's further report, N.W.P. and Oudh, Proceedings for December 1897 (No. 4). RAU, p 2
Political Department (1897g) Government of the N.W.P. and Oudh to the commissioner of Kumaon, communicating orders on the subject of the dues levied by Thibetans from British subjects, 6 December, N.W.P. and Oudh, Proceedings for December 1897 (No. 13). RAU, pp 6–7
Political Department (1897h) Deputy commissioner of Almora to commissioner, Kumaon division, 29 July, N.W.P. and Oudh, Proceedings for August 1897 (No. 38). RAU, p 47

Political Department (1897i) Government of the N.W.P. and Oudh to the commissioner of Kumaon, intimating, with reference to serial no. 76, that the lieutenant-governor was not prepared at present to confer revenue and magisterial powers on the peshkar at Garbyang; and sanctioning the proposal to make permanent the political establishment entertained at that place, which had hitherto been treated as temporary, 9 December, N.W.P. and Oudh, Proceedings for December 1897 (No. 14). RAU, p 7

Raj K (2002) When human travelers become instruments: the Indo-British exploration of Central Asia in the nineteenth century. In: Bourguet MN, Licoppe C, Sibum HO (eds) Instruments, travel and science: itineraries of precision from the seventeenth to the twentieth century. Routledge, London, pp 156–188

Rajan R (2006) Modernizing nature: forestry and imperial eco-development 1800–1950. Oxford University Press, Oxford

Ramble C (1993) The name Bhotey. Himalayan 6(5):17

Ramble C (1997) Tibetan pride of place: or, why Nepal's Bhotiyas are not an ethnic group. In: Gellner DN, Pfaff-Czarnecka J, Whelpton J (eds) Nationalism and ethnicity in a Hindu kingdom. Harwood Academic Publishers, Amsterdam, pp 349–414

Ramble C (1998) The classification of territorial divinities in Pagan and Buddhist rituals of south Mustang. In: Blondeau AM (ed) Tibetan mountain deities, their cults and representations. Verlag der Österreichischen Akademie der Wissenschaften, Wien, pp 123–143

Ramble C, McKay A (2003) Se: preliminary notes on the distribution of an ethnonym in Tibet and Nepal. In: McKay A (ed) The history of Tibet, vol I, The early period to c. AD 850, the Yarlung dynasty. Routledge Curzon, London, pp 69–89

Raper FV (1812) Narrative of a survey for the purpose of discovering the source of the Ganges. Asiat Res 11:446–564

Rawat DS, Sharma S (1997) The development of a road network and its impact on the growth of infrastructure. A study of Almora district in the central Himalaya. Mt Res Dev 17:117–126

Rizvi J (1999) The trade in pashm and its impact on Ladakh's history. In: van Beek M, Bertelsen KB, Pedersen P (eds) Ladakh: culture, history and development between Himalaya and Karakoram. Aarhus University Press, Aarhus, pp 317–338

Roy T (2003) Changes in wool production and usage in colonial India. Mod Asian Stud 37(2):257–286

Sahlins M (1985) Islands of history. University of Chicago Press, Chicago

Sahlins M (2004) Apologies to Thucydides: understanding history as culture and vice versa. University of Chicago Press, Chicago

Sahlins M (2010) The whole is a part: intercultural politics of order and change. In: Otto T, Bubandt N (eds) Experiments in holism: theory and practice in contemporary anthropology. Wiley, Malden, pp 102–126

Samuel G (1993) Civilized shamans: Buddhism in Tibetan societies. Smithsonian Institution Press, Washington, DC

Sanwal RD (1976) Social stratification in rural Kumaon. Oxford University Press, New Delhi

Sax WS (1991) Mountain goddess: gender and politics in a Himalayan pilgrimage. Oxford University Press, New York

Sax WS (2009) God of justice: ritual healing and social justice in the central Himalayas. Oxford University Press, New York

Sax WS (2010) The royal pilgrimage of the goddess Nanda. Scr Inst Donneriani Aboensis 22:334–352

Schrader H (1988) Trading patterns in the Nepal Himalayas. Breitenbach, Saarbrücken

Scott JC (2009) The art of not being governed. An anarchist history of upland Southeast Asia. Yale University Press, New Haven

Sherring CA (1974 [1916]) Western Tibet and the British borderland. The sacred country of Hindus and Buddhists. With an account of the government, religion and customs of its peoples. Cosmo Publications, Delhi

Shneiderman S (2010) Are the central Himalayas in zomia? Some scholarly and political considerations across space and time. J Glob Hist 5(2):289–312
Shneiderman S (2013) Himalayan border citizens: sovereignty and mobility in the Nepal-Tibetan Autonomous Region (TAR) of China border zone. Polit Geogr 35:25–36
Shneiderman S (2015) Rituals of ethnicity: Thangmi identities between Nepal and India. University of Pennsylvania Press, Philadelphia
Srivastava RP (1966) Tribe-caste mobility in India and the case of Kumaon Bhotias. In: von Fürer-Haimendorf C (ed) Caste and kin in Nepal, India and Ceylon. Asia Publishing House, Bombay, pp 161–212
Stewart GT (2009) Journeys to empire: enlightenment, imperialism, and the British encounter with Tibet, 1774–1904. Cambridge University Press, Cambridge
Tambiah SJ (1977) The galactic polity: the structure of traditional kingdoms in Southeast Asia. Ann N Y Acad Sci 293:66–97
Teltscher K (2006) The high road to China: George Bogle, the Panchen Lama and the first British expedition to Tibet. Bloomsbury, London
Tolia RS (2009) Founders of modern administration in Uttarakhand: 1815–1884. Edward Gardner to Henry Ramsey. Bishen Singh Mahendra Pal Singh, Dehra Dun
Traill GW (1992 [1828]) Statistical sketch of Kumaon. In: Joshi MP, Fanger AC, Brown CW (eds) Himalaya: past and present, vol II. Shree Almora Book Depot, Almora, pp 1–97.
Traill GW (1992 [1832]) Statistical report on the Bhotiya mehals of Kumaon. In: Joshi MP, Fanger AC, Brown CW (eds) Himalaya: past and present, vol II. Shree Almora Book Depot, Almora, pp 99–154
Trivedi GM (1991) Descriptive grammar of Byansi – a Bhotiya language. Anthropological Survey of India, Calcutta
Turin M (2002) Ethnonyms and other-nyms: linguistic anthropology among the Thangmi of Nepal. In: Buffetrille K, Diemberger H (eds) Territory and identity in Tibet and the Himalayas. Brill, Leiden, pp 253–270
Turner S (1800) An account of an embassy to the court of Teshoo lama, in Tibet; containing a narrative of a journey through Bootan, and part of Tibet. To which are added, views taken on the spot, by Lt. Samuel Davis; and observations botanical, mineralogical, and medical, by Mr. Robert Saunders. Printed by W. Bulmer and Sold by G. & W. Nicol., London
van Schendel W (2002) Geographies of knowing, geographies of ignorance: jumping scale in Southeast Asia. Environ Plan D Soc Space 20(6):647–668
van Schendel W (2007) The Wagah syndrom: territorial roots of contemporary violence in South Asia. In: Basu A, Roy S (eds) Violence and democracy in India. Seagull Books, Oxford, pp 36–82
van Spengen W (1998) On the geographical and material contextuality of Tibetan pilgrimage. In: McKay A (ed) Pilgrimage in Tibet. Curzon Press, Richmond, pp 35–51
van Spengen W (2000) Tibetan border worlds. A geohistorical analysis of trade and traders. Kegan Paul International, London
Vasan S (2006) Indo-Tibetan border trade in Himachal Pradesh. China Rep 42(1):41–56
Vitali R (2003) A chronology (bstan rtsis) of events in the history of Mnga'ris Skor Gsum (tenth-fifteenth centuries). In: McKay A (ed) The history of Tibet, vol II. The medieval period, c. 850–1895, the development of Buddhist paramountcy. Routledge Curzon, London, pp 53–89
von Fürer-Haimendorf C (1964) The Sherpas of Nepal. John Murray, London
von Fürer-Haimendorf C (1975) Himalayan traders: life in highland Nepal. Cox and Wyman Ltd., London
Waller D (1990) The pundits: British exploration of Tibet and Central Asia. University of Kentucky Press, Lexington
Walton HG (1911) Almora: a gazetteer, vol XXXV of the district gazetteers of the United Provinces of Agra and Oudh. Government Press United Provinces, Allahabad
Warikoo K (1990) Ladakh's trade relations with Tibet under the Dogras. China Rep 26(2):133–144

Whalley P (1991 [1870]) British Kumaon. The law of the extra-regulation tracts subordinate to the government, N.W.P. Vishwavidyalaya Prakashan, Varanasi
Willis CM (2007) A descriptive grammar of Darma: an endangered Tibeto-Burman language. University of Texas, Austin, PhD-dissertation
Younghusband F, MacDonald R, Gore J, Freshfield D (1905) Exploration and survey with the Tibet frontier commission, and from Gyangtse to Simla via Gartok: discussion. Geogr J 26:391–394
Yule H (1877) Presentation of the royal and other awards (at the anniversary meeting, May 28). Award of medals – Nain Singh. J R Geogr Soc 47:cxxv–cxxvii

Chapter 3
Tribal Identities and Scalar Politics in Postcolonial India

Abstract In this chapter I analyze how the Bhotiyas have leveraged the performativity of scale through various narratives and social practices to effectively voice their demands for tribal status in postcolonial India. The argument is divided into four empirical parts: As an entry point into the discussion, the first part introduces the narrative of Old Lady Jasuli, a widow who lived in the Darma valley during late colonial rule. The activities of the Kumaon Bhotiya Peoples' Federation (KBPF), which was founded shortly before India's independence in 1947, are analyzed in the second part. Its' members re-assembled representations that the British had coined about Kumaon's trans-Himalayan traders, especially in terms of gender relations. In so doing they articulated the Bhotiyas' tribalness by producing *Bhot* (or the region inhabited by the Bhotiyas) as a scale effect in independent India. This effect of scale materialized as a 'real' frame of social action when state officers notified the Bhotiyas as a Scheduled Tribe (ST) in 1967. Only a few years later, as explored in the third part, a new generation of politically organized Bhotiyas started to challenge and transform these inherited scalar categorizations and arrangements, which were then perceived to reinforce rather unequal power relations within the ST community. They deployed the character of Old Lady Jasuli to convey an alternative sense of their tribal identity whilst reproducing independent India as a scale effect in *Bhot*. In the fourth part I then analyze how caste Hindus challenged the Bhotiyas' tribal status when Uttarakhand was carved out as an independent state of the Indian Union in 2000. These new competitors staked the claim that the Bhotiyas' ST-designation is a scalar expression that does not pertain to a specific people but to a region, including all its 'native' residents. The research contributes to emergent scholarship focusing on the ways in which identity and scale interact as categories of practice, particularly at times of macro-institutional transformations: the shift from British imperial rule to nation-state sovereignty or the creation of new states within a country's federal system of government.

Keywords Ethnic identification • Scalar politics • Performativity of scale • Scheduled Tribes • Bhotiyas • Kumaon Himalaya • Postcolonial India • Macro-institutional transformations

C. Bergmann, *The Himalayan Border Region*, Advances in Asian Human-Environmental Research, DOI 10.1007/978-3-319-29707-1_3

On 17th April 1947, about 4 months before India won independence from the British Empire, the newly founded Kumaon Bhotiya Peoples' Federation (KBPF) submitted before the Indian Minority Sub-Committee of the Constituent Assembly's Advisory Committee a report that demanded special statutory safeguards as well as constitutional protection for the Bhotiya community as a minority. The Federation consisted of a powerful elite that had already amassed a considerable amount of cultural capital under British rule, especially by having accessed educational institutions and offices in the public sector. Mr. K.S. Pangtey (M.A., B.Com, LL.B) from Johar, for instance, who is honored in the report to have drafted the English version, was the Bhotiyas' first postgraduate (KBPF 1947: 2). Under the heading 'Culture of the Bhotiyas' one finds the following entry:

> 'Culture is that complex whole which includes knowledge, belief, art, morals, law, custom, and any other capabilities and habits acquired by man as a member of society' defines Taylor [sic]. In other words, culture in human society is a collective learning process produced by the counter action of the individual minds. The organic evolution furnishes the capacity. So the culture in all its forms whether tool making, language, art, religion, morality or Govt. [sic] has its organic basis of transmission and development, the long physical and social infancy of the human individual. In short culture is the means to the ends of life. The ends are rooted in our natural inheritance.
>
> Taking into consideration these fundamentals of culture it can be observed that the Bhotiyas of Kumaon are still in the lower rungs of the cultural evolution. It is because of the fact that environment is still dominant with its strong impress not only upon their social economy but also upon their religious ceremonies, festivals and social practices. But there is no denying the fact that tremendous difficulties of environment have not, as it might be imagined, been able to suppress the initiative and energy of the Bhotiyas. On the other hand they have possessed a high degree of adaptive skill and uncommon physical vigour. But it still remains to be said as Golden-Weiser [sic], the great anthropologist has rightly pointed out that 'as we pass from primitive to more civilized conditions, the dependence of material culture of a group upon its physical environment becomes less and less conspicuous'. Their age-old methods of trade and barter; crude agricultural implements; simple and primitive looms; their superstitions and beliefs; habits and social practices and above all their nomadism; all go to show that they have only emerged from the primitive ways of life. Being a trader he comes in contact with the great currents of thought and feelings, unlike more stay-at-home mountaineers, but he has little leisure to study and so to practice them.
>
> Charles A. Sherring commenting on the religion of the borderland Bhotiyas in his book 'Western Tibet and the British Borderland' writes 'In my opinion this fact can only be accounted for on the theory that the Bhotiyas left Tibet before writing was introduced into that country about 650 A.D. and that the physical difficulties surrounding their homes in the giant mountains of the Himalayas were so great that they had been left, ever since untouched by the influences at work on both sides of them viz. in Tibet and India. It is nothing but their seclusion among these vast mountains that can account for the fact that they have among them none of the doctrines of Buddhism, or the common practices of Lamaism, and Hinduism has been till quite recently unknown and that their own language has never been reduced to writing. Further I consider it wonderful and I think the reader will agree with me as he reads their customs and habits that we can find among them so little of the degrading immorality and demon worship of the Bon faith, which must have been in vogue in Tibet before the time of their immigration and it speaks of their natural character that although they have been for centuries in constant touch with Tibet for business purposes, they have never absorbed into their own simple religion the extravagances and demonology of Lamaism'.

> At the present moment the Bhotiyas of Garhwal, Johar, Darma, Bians [sic] and Chaudas [sic] have almost been Hinduised, though some of them have their own tribal religion and social practice. It was on account of these facts that Missionaries many years before tried to convert them to Christian faith, but it was vehemently resisted. They better preferred to adopt the Hindu faith of the socially higher groups of Kumaon of their own accord.
>
> So we find that the Bhotiyas have a culture of their own.
>
> Sherrings interpretation that the Bhotiyas have first migrated from Tibet in the dim past is perhaps right, because of the fact that the Tibetan feudal chiefs considered the whole of Bhotiya region under their rule for a long time and still exercise this right in Mana, Garhwal, by way of imposing tax and in Bians and Chaudans not only by imposing certain taxes but also by taking forced labour or Begar. 'The loft patties of Darma, Bians and Chaudans nominally form part of the British Empire, our geographical boundary with Nari Khorsum or Hundesh (Great Tibet) being the main Himalayan chain forming the watershed between the two countries. In spite of this actual territorial right I found at the time of my visit in 1897 that it was impossible not to agree with the notions in asserting that British prestige and protection in those regions were mere myths; that Tibetan influence alone was dominant and prevailing and Tibetan Law enforced and feared. The Tibetans in fact openly claim possession of Pattis bordering on Nari Khorsum …' writes Mr. A Henry Savage Landor in his book 'In the Forbidden Land'. Even after 50 years, these facts are still true to some extent though it is fast diminishing. Again the existence of Mane Walls, certain social dietic habits are perhaps the best vestiges of our Tibetan heritage. This is how we were called to be Bhotiyas, though some of us resent to be associated with the dirty Tibetans as the civilized man of today with monkeys and Homo Sapiens. Locally we are known as Shokas [sic]. (KBPF 1947: 13–14)

Before the report was eventually submitted, the KBPF had sent some representatives to Delhi. Dr. B.R. Ambedkar, a Dalit member of the Constituent Assembly who became independent India's first law minister, looked into their matter and made suggestions on how to improve the draft for reaching at an even more appropriate and convincing form (Pangtey 1992: 65–68).[1] The quoted passage, like many others throughout the whole report, is intriguing. Following Pratt (1991: 35), it can be seen as an 'autoethnographic text', or "a text in which people undertake to describe themselves in ways that engage with representations others have made of them". It forms the initiation of a dialogue with the political leaders of the awakening Indian nation, leaders who increasingly took up a position for the 'weak' sections of society. On 14th August 1947, 1 day before coming into office as independent India's first Prime Minister, Jawaharlal Nehru delivered his momentous speech 'Tryst with Destiny' to the Constituent Assembly. There he announced the dawn of national freedom and emphasized that "service to India means the service to the millions who suffer. It means the ending of poverty and ignorance and disease and inequality of opportunity" (Nehru 2007 [1947]: 328). The Federation clearly rated its people among the side of the suffering. The reduction in trans-Himalayan trade had resulted in a loss of livelihood opportunities and the process of nation-building

---

[1] From 1913 onwards Dr. Ambedkar had studied at Columbia University, USA. After his graduation in 1916 he delivered a paper in a seminar held by the anthropologist A.A. Goldenweiser, to whom the Federation also referred in its report. This paper was published in the very same year under the title *Castes in India: Their Mechanism, Genesis and Development* (Ambedkar 1917).

was estimated to possibly aggravate the traders' political, economic, geographical and, especially, cultural marginalization.

For getting this point across to the Indian Minority Sub-Committee, the report's authors skillfully integrated into their argument colonial appraisals on their community's ethnogenesis. Charles S. Sherring (1974 [1916]: 69), a former deputy commissioner of Kumaon, produced one of the most authoritative sources in that regard. He argued that the Bhotiyas' ancestors had emigrated from Tibet in the ancient past, therewith synthesizing a debate that existed for as long as British rule in the area. Administrators found it rather difficult to determine the ethnic origin of the communities involved in trans-Himalayan trade. They were particularly bewildered by their Tibetan-like appearance and the rather fluid boundaries of belonging that seemed to characterize everyday life on the ground (Brown 1992). It should therefore not come as surprise that numerous alternative speculations surfaced in official documents. One of the most influential was already introduced by Traill (1992 [1832]: 115), the second British commissioner of Kumaon, who considered the inhabitants of the Darma valley as a special case that would not disprove but rather qualify the hypothesis of Tibetan ancestry among the Bhotiyas.

The Darmani would differ from other Bhotiya groups both in terms of their facial appearance and cultural outlook, and even their dress seemed to be knit in quite another fashion. Traill (1992 [1832]: 115), therefore, purported that they were descendents of a horde of warriors who had presumably entered Kumaon from the Indo-Gangetic plain in the sixteenth century. Plagued by diseases these warrior-nomads were eventually reduced to a very small number and – concerned for their future existence – found a safe refuge in the remote heights of the Darma valley. He explicitly associated this horde with the powerful ruler Timur, whose armies had conquered parts of West-, South-, and Central Asia two centuries before in attempting to lead the Mongol Empire (Tatary) to new heights. This hypothesis, then, amounts to the following picture: all Bhotiyas were supposed to be of a Mongol Tatar origin, but their ethnogenesis was thought to differ in that the Darmani cannot be traced directly from Tibet but rather "through the circuitous route of Hindustan" (Traill 1992 [1832]: 115).

The evidence that bureaucrats of the colonial state provided for supporting such hypotheses was undeniably rather scant, highly disputed and also disliked by the natives (Traill 1992 [1832] 116; cf. Landor 1905: 222–223). Moreover, such hypotheses also reflected and reinforced power and interest constellations amongst the Bhotiyas themselves. As described in the previous chapter, members of the Rawat-clan from the Johar valley established a rather close relationship with the British colonial administration, and this proximity also influenced British perceptions. Sherring (1974 [1916]: 64), for instance, stated openly that the Johari looked down upon their eastern neighbors, especially the Darmani, for their assumed primitiveness, and he perpetuated associated stereotypes as an ethnographic truth. Least influenced by outside forces, the Darmani would be trapped in a snare of naïve superstitions and disreputable customs. Between the lines, however, it becomes clear that this assessment also had deeper political roots. The eastern Bhotiyas were known to have maintained stronger ties of loyalty and obligation to Nepalese rulers than to the kings of Kumaon (Sherring 1974 [1916]: 69). While the Johari had been able to sustain their influence across the regime transitions from Chand to Gorkha and from Gorkha to British rule, the eastern Bhotiyas found it more difficult to capitalize on their alliances and tributary relations in the political arena of the late colonial state.

Eventually, however, British administrators applied a classificatory scheme that seemed less arbitrary and was independent of any local nomenclature. This was achieved by characterizing the Bhotiyas as a 'Mongoloid border tribe' (Atkinson 1973 [1882]: 111–115; cf. Srivastava 1966: 204).[2] Whilst the designation 'Mongoloid' was a widely used racial category of the time, the affix 'border tribe' articulated the perceived differences of the people so classified. On the one hand the Bhotiyas had long been identified in terms of their location on the imperial margins, namely as residents of a *border* region that the British had (mistakenly) recognized as *Bhot* (Brown 1992). On the other hand the transregional scope of their livelihoods should not hide the fact that these itinerant traders would lead the rather 'primitive' life of a *tribal* people. Having thus already been legally registered as 'tribe' in colonial times, the Federation's claim to the status of a national minority appeared instantly warrantable.

Due to an opinion articulated by the British adventurer Arnold H.S. Landor (1899: chapter 7), this claim received even more political weight. Landor pointed out that Tibetan rule was strongly felt in the area and that local residents suffered as a result of their neglect by the British Empire. The KBPF was quick to add that the Tibetans continued to wield influence in the region. By granting the Bhotiyas special statutory safeguards and constitutional protection, the Indian Minority Sub-Committee could thus not only achieve its own aims relative to minorities, but also foster the territorial integrity of the Indian nation state in a sensitive border area. This way of depicting things also helped the Federation to accentuate their community's resistence to the 'foreign' influences of both Lamaism and Christianity. At the same time some community members had, albeit with limited success, attempted to promote Hindu cultural practices from other regional groups with a higher social standing. The overall image assembled thus results in what Ghurye (1943) called 'primitive Hindus', which was skillfully substantiated with reference to the work of two influential anthropologists.

Edward B. Tylor's (1871: 1) definition of culture, which every anthropology student still learns at the beginning of her study, is considered to be the first explicit one within the discipline (Kroeber and Kluckhohn 1952). It stems from the opening page of a book entitled 'Primitive Culture'. Whether intentionally or not, the quotation provided in the report elides the second and third words of the original sentence, starting with 'culture *or civilization*'. These two terms, however, are important to understand Tylor's position. For him culture describes a singular and universal phenomenon, something that every human being has, to varying degrees, within one overriding and continuous evolutionary process (Barnard and Spencer 1996: 138).

[2] Even today the English term 'Mongoloid' is widely used in local discourses. For a discussion of this circumstance see Nawa (2000).

The variables at work within this process are not racial features but the cultural inventories that are generated by humans in their society and under specific environmental conditions. Although even the most 'savage' tribe would in principle be capable of advancing toward the higher stages of civilization, the developmental forms envisaged by Tylor are nevertheless of a sequential order and driven by the rather mechanical spread of cultural traits. But the Federation argued "that the Bhotiyas have a culture of their own" (KBPF 1947: 14), thereby escaping Tylor's unilinear framework by claiming cultural uniqueness. Its members did not want to dismiss, but rather receive the means to uplift their unique tribal culture as they saw fit.

The reference to Alexander A. Goldenweiser (1916: 629) was chosen by the authors of the report to give this latter point momentum. Goldenweiser was one of Franz Boas' favorite students who followed his teacher in remaining critical of the evolutionary perspective. According to him there is not one culture or civilization but many. They differ in terms of achievement and variety but are still "on a par and comparable" (Goldenweiser 1924: 130). While the scope of cultural variation is limited by the natural environment, particularly among the so-called 'primitives', Goldenweiser considered the fate of such communities much more dependent on the dominance and hegemony of the modern Western world. In his view, this overwhelming force makes indigenous groups "lose interest in their own ideas" (Goldenweiser 1924: 131). Too weak to resist they would not only be prone to capitulate but might even disappear altogether in a world of nation state politics.

However, the Federation withheld Goldenweiser's 'sentimental pessimism' (Sahlins 1999: i) and instead followed a more optimistic Boasian spirit according to which 'primitive cultures' are equipped with improvisational capacities to invent, imitate or acculturate new forms of action and therewith adapt to changing framework conditions. The crux of the matter is that marginalized groups might not always be able to achieve this on their own. As Nehru (1973) later put it, India's tribal populations are entitled for benefits on offer from the state because only in that manner would it be possible for them to develop in ways that resonate with their own 'genius'.[3] But it took another 20 years after the Federation had submitted its report to the Constituent's Assembly Advisory Committee until the Bhotiyas were officially notified as a Scheduled Tribe (ST) and thereby made eligible to receive the envisaged preferential treatment from the Government of India. In the interim their negotiations took many twists and turns and, indeed, they have yet to come to a final end.

[3] Boas (1896: 5) had referred to the "genius of the people" in explaining how certain cultural elements, such as myths, become adopted and changed by cultural groups. Based on this conception he reconstructed the interrelations between different cultures and highlighted the creativity through which the diffusion of specific cultural traits is driven (Boas 1937).

3.1 The Notion of 'Tribe'

The notion of 'tribe' has a long and exceptional career, both in academic and even much longer in politico-administrative circles.[4] Within the Roman Empire *tribus* denoted the constantly growing division of citizens (van Schendel 2011: 19). Centuries later the European colonial powers adopted it as a classificatory device for ranking local cultures into the fold of an "all-embracing civilization" (Spurr 1993: 197). In South Asia colonial administrators initially used the terms of 'caste' and 'tribe' more or less interchangeably and only distinguished them when the broadening scope of censuses, surveys and gazetteers required a refined categorization (Cohn 1987: 224–254).[5] Various criteria were then devised to identify tribal populations, but a clear-cut definition of the category proved elusive, and remains so to the present day (Xaxa 1999; Pfeffer and Behera 2002; Rycroft and Dasgupta 2011).

The tribal situation emerged as an integral component of India's national agenda (van Schendel and Bal 2002: 126). The Constitution of India guarantees the equality of all citizens before the law (Article 14). It therefore prohibits any discrimination on the grounds of religion, race, caste, sex or place of birth (Article 15) and mandates equality of opportunity in matters of public employment (Article 16). In order to ensure these fundamental rights to underprivileged communities, the constitution further specifies formal mechanisms to identify and schedule castes (Article 341) and tribes (Article 342) that are in need of preferential treatment so as to catch up and successfully integrate with mainstream society.[6] During the Constituency Assembly debates, Dr. B.R. Ambedker (CAD 1949: 1637) explained that the purpose of including these latter articles

> was to eliminate the necessity of burdening the Constitution with long lists of Scheduled Castes and Scheduled Tribes. It is now proposed that the President in consultation with the Governor or Ruler of a State should have the power to issue a general notification in the Gazette specifying all the Castes and Tribes or groups thereof deemed to be Scheduled Castes and Scheduled Tribes for the purposes of the privileges which have been defined for them in the Constitution.

---

[4] For the mutual impact of administrative and academic conceptions of 'tribe' see Sinha (2004) and Baviskar (2005: 5106).

[5] The most influential of the British anthropologist-administrators in the Indian civil service was H.H. Risley. He played a major role in establishing anthropometry as a scientific foundation for identifying, describing and comparing racial or ethnic formations (Bates 1995). For scholarly analyses of the 'ethnographic' state in India see Pels and Salemink (1999), Bayly (1999), Dirks (2001) and Peabody (2001).

[6] Article 342 of the Indian Constitution reads as follows (CoI 2011): "Scheduled Tribes. (1) The President may with respect to any State or Union territory, and where it is a State, after consultation with the Governor thereof, by public notification, specify the tribes or tribal communities or parts of or groups within tribes or tribal communities which shall for the purposes of this Constitution be deemed to be Scheduled Tribes in relation to that State or Union territory, as the case may be. (2) Parliament may by law include in or exclude from the list of Scheduled Tribes specified in a notification issued under clause (1) any tribe or tribal community or part of or group within any tribe or tribal community, but save as aforesaid a notification issued under the said clause shall not be varied by any subsequent notification".

In other words, the president is meant to consult with the political leaders of each state for their recommendation on the groups entitled for a public notification as Scheduled Caste or Scheduled Tribe. The final decisional power, however, rests exclusively with the president. In order to make the overall procedure more transparent, the newly appointed commissioner for Scheduled Castes and Scheduled Tribes devised various test indicators in 1952 (RoC 1952: 11). For Scheduled Tribes these included a tribal origin, a 'primitive' way of life, residence in a remote habitation and shyness of contact with the society at large, as well as a general backwardness in all respects. It is easy to see that these indicators, which are in use until today, leave plenty of room for interpretation. The whole policy thus largely depends on both the manner in which possible candidates promote themselves and on the judgments and calculations of state officials and government bureaucrats who validate their performance and advise the president.

There are numerous examples in which people of the 'same' group were included in the list of Scheduled Tribes in one federal state of the Indian Union but excluded in another (Kapila 2008). Moreover, some states refused in principle to include the tribes whom the British had classified as 'criminal' or 'denotified', while others did not. And because India's neighbors implemented different policies, further ambiguities arose for migratory groups, such as the Bhotiyas, whose residential patterns cut across national or federal state borders. The name 'Bhotia' was already included in the Constitution (Scheduled Tribes) Order of 1950. However, public notification was delayed for political reasons. Until 1967 the Government of Uttar Pradesh, the state to which the territory of present-day Uttarakhand had been aggregated after India's national independence, had not submitted a corresponding recommendation. By then, however, the international crisis between India and China as well as a rising political activism in the Garhwal and Kumaon Himalaya had piled the pressure for more generous concessions on behalf of the Bhotiyas and other tribal groups in the region.

Through measures of 'affirmative action', including reserved seats in the legislature, public service and educational institutions, the Government of India seeks to dissolve the scheduled groups' 'weakness' (Shah and Shneiderman 2013). Although Scheduled Castes and Scheduled Tribes compete for similar provisions, the general policy differs for both target groups. Whereas the former are supported in that they might eventually become full-fledged mainstream citizens, tribal communities are scheduled not only to bring them into contact with mainstream society, but also to preserve their cultural distinctiveness and relative autonomy as *adivasis* (Galanter 1984; Corbridge 2000).[7] One major policy instrument in this regard is the

[7]The term *adivasi* already gained popularity in pre-independent India and derives from Sanskrit *adi* (meaning 'beginning' or 'of earliest times') and *vasi* (meaning 'residents of'). It is thus close in meaning to the English term 'aborigine' and until today widely used in academic writings on India's tribes. South Asia's marginal groups, however, have recently begun to adopt the transnational concept of 'indigenous peoples' for themselves (Karlsson 2003; Karlsson and Subba 2006). According to van Schendel (2011: 25–27) this concept turns the idea of tribes on its head: it is not based on assumptions of backwardness, wildness and isolation from the mainstream but rather marginalizes the mainstream – as not being native or autochthonous – and therewith challenges the

specification of Scheduled Areas, of territorial units whose tribal residents receive special financial support for fostering a self-determined developmental path.[8] In order to move out of poverty and up the social ladder, tribal communities are not only to be developed – like Scheduled Castes – but also protected from outside interventions "that seek to exploit their ignorance and innocence" (Corbridge 2000: 69).[9]

Policy makers and development practitioners still often equate tribal distinctiveness with ignorance and naivety, factors that are then made responsible for tribal groups' vulnerability when being drawn out of their alleged isolation. An entry in India's Tenth Five Year Plan (2002–2007), for instance, points out that

> [t]ribal communities continue to be vulnerable even today, not because they are poor, assetless and illiterate compared to the general population; but often their distinct vulnerability arises from their inability to negotiate and cope with the consequences of their integration with the mainstream economy, society, cultural and political systems, from all of which they were historically protected by their relative isolation. (GoI 2001: 457)

Without denying the suffering that marginalized tribal groups often experience, I doubt the general validity of such claims and rather suggest that observers have often failed to recognize the manner in which these communities negotiate the consequences of outside interference. Sahlins (1999: ix; cf. Sahlins 1993) coined the phrase 'indigenization of modernity' to grasp the resourcefulness and agency of kinship-based societies in adapting new inputs and demands to meet their needs. Up to the present there are hunter and gatherers who continue to hunt and gather, foragers who forage, and nomadic peoples who move over long distances to seek grazing for their flocks. However, the fact that such groups continue with their traditional livelihood practices should neither be taken as an indicator of pristine authenticity nor of stasis. On the contrary, they often uphold identities that challenge modernist agendas and simultaneously harness the availability of up-to-date technologies and categorical inventories to their seemingly 'paleolithic' purposes. In other words, marginalized people are "tactically selective about modernity" and the tactics they chose assign specific 'signatures' to broader processes of change and development

---

time honored dichotomy between civilization and barbarism. It allows people to self-confidentially claim a kind of global citizenship, which – in some cases – might even encourage them to seek independence from the state to which they belong. The official term to denote Scheduled Tribes in Hindi is *anusuchit janjati*. For further terminologies applied to such groups in the South Asian context see van Schendel and Bal (2002).

[8] Scheduled Areas are autonomous and federally administered areas within the state. They are normally compact in size, under-developed, and populated by one Scheduled Tribe. In Uttarakhand six blocks – including Dharchula and Munsiari – were proposed to the Ministry of Tribal Affairs for being classified as Scheduled Areas in 2001 (UHC 2004: 3).

[9] Because the Constitution of India institutionalizes a pluralistic self-image of Indian society, some authors have referred to it as a kind of multi-culturalist text (Kapila 2008: 120; cf. Bhargava et al. 1999; Rudolph and Rudolph 2002). For the contemporary debate on multiculturalism and the associated politics of recognition consult Taylor (1994), Bauman (2001), Benhabib (2002), Fraser and Honneth (2003).

(Michaud 2012: 1854).[10] Moreover, even those who become engaged in non-traditional occupations may often refrain from discarding their traditional ties and practices as obsolete, or even as harmful, and instead develop strategies to keep them alive and thrilling (Fisher 2001).

This means that economically and politically weak societies are not only changed by outside forces but are in many cases also perfectly able "to interpret, adapt and even subvert resulting pressures" (Michaud 2012: 1854–1855). While indigenization concerns the manner in which marginalized people 'digest' the higher-level dynamics that are affecting their local-scale control over development, I am concerned in this chapter with the broader dynamics of a contentious scalar politics. Put simple, scales are commonly used spatial representations such as the local, the national and the global. The most prominent metaphors applied to explain scales are 'scale as size' (e.g. map scale) and 'scale as level' (e.g. a pyramid or hierarchy of scales) (Howitt 1998). However, scholars increasingly argue that these metaphors are misleading because they suggest that scale sets up an external framework to which people react "rather than a strategy that they mobilize to realize their interests" (Purcell and Brown 2005: 281) and an effect that they produce in the course of their practical engagements (Kaiser and Nikiforova 2008).

Two interrelated sets of questions, which I tackle in this chapter, present themselves here: First, there is the issue of how members of the KBPF inserted a contentious scalar politics into their demands for special statutory safeguards and constitutional protection. What lines of action did they take to convincingly articulate these demands toward decision makers at 'higher' levels of activity, specifically at the national stage? And what effect of scale did their strategic maneuverings produce? The second set of questions has to do with the strategies of Bhotiyas, who, after their community had been notified as a Scheduled Tribe in 1967, began to challenge and transform the power relations of inherited scalar categorizations and arrangements. How did they re-envision the ties between the 'lower' scale of a remote mountain region that formed the material conditions of their tribal existence and the broader currents of societal change in India? What consequences did this re-envisioning have in terms of practical strategy and political contestation? Before going into ethnographic detail, a few words on the rich body of anthropological literature on ethnicity and its relationship to the core geographical concept of scale is due.

3.2 Ethnic Identities and Scalar Politics

The anthropological debate on ethnicity received a major impetus through the work of Fredrik Barth (1969; cf. Jenkins 2008: chapter 2; Wimmer 2013: 22–23). Barth directed attention away from the hitherto axiomatic view that geographical and

[10] Such a perspective resembles the literature on 'alternative modernities', which questions the teleological aspect of modernization theory and suggests instead that each 'modernizing' region must be understood on its own merits (Gaonkar 2001; Kaviraj 2000).

social isolation were the principle reasons for ethnic diversity (Kreutzmann 2006: 51). Instead of searching for some particular cultural traits that define discontinuous communities, Barth focuses on the socio-cultural processes of inclusion and exclusion that enable actors to maintain ethnic boundaries despite or even because of their constant interaction with other groups. The lesson from Barth, who more or less defined the mainstream anthropological perspective on ethnicity over several decades, is that we should not look for constellations of traits but rather focus on the strategies through which people render ethnic identities meaningful and visible in everyday practice:

> It is important to recognize that although ethnic categories take cultural differences into account, we can assume no simple one-to-one relationship between ethnic units and cultural similarities and differences. The features that are taken into account are not the sum of 'objective' differences, but only those which the actors themselves regard as significant. (Barth 1969: 14)

The standards of evaluation according to which people arrive at judgments of ethnic difference and hierarchy are not random but follow from certain contextual logics (Barth 1969: 27). Barth exemplifies this with reference to the Pathan (Pasthun) borderland between Afghanistan and Pakistan. Throughout the region, claims of ethnic-political leadership are evaluated according to an actor's ability to provide guests with hospitality and gifts. But for some influential resident groups, like the Kohistanis, hospitality could not provide a basis for assuming a superior position due to their comparatively low economic standard of living. However, by emphasizing other aspects of a distinguished Kohistani ethnicity these people scored "quite high *on the scales that then become relevant*" (Barth 1969: 25, italics mine). In the localities where they resided it was the public display of their autonomy and independence as landholder-cultivators that positioned them as political equals and allies to the more affluent Pathans.

While Barth explicitly adopts the term 'scale' in his analyses of ethnic boundary making, he more or less uses it as a synonym for context. But he refined his thinking about scale in the introduction to another edited volume entitled 'Scale and Social Organization' (Barth 1978a). Although this work received far less public attention than his previous writings on ethnicity, anthropologists have recently begun to acknowledge its relevance to contemporary discussions on scale (Schlee 2011). Barth turns against all those who consider scale as a taken-for-granted concept and sell vacuous generalizations as revealing scientific insights. By way of example, what does it tell us that the features of small-*scale* societies are conditioned by their small size while those of large-*scale* societies originate from their largeness? In order to arrive at some more substantial assertions he asks his readers to contemplate a single but concrete question:

> [W]e recognize that a neighborhood with 50 residents in the Andaman islands, a Thai village, and a New York City differ not just in features such as kinship system, religious organization, or forms of property relations – they are fundamentally differently constituted internally and in external relations in ways which entail different orders of magnitude of size, numbers and complexity: i.e. different *scale*. [...] How do we transform this recognition – not by producing another framework for small-scale/large-scale, but by constructing a framework for discovering what in fact are the basic differences in the constitution of men and societies? (Barth 1978a: 10–11)

Scale is defined here in relational terms, namely as the interplay of size, number and complexity at different levels of organizational order. Based on this working definition, Barth draws attention to the fact that scale is not an arbitrary artifact of our analyses but rather something whose specific form of appearance needs to be established empirically. In other words, one should not assume beforehand that the world of the Andaman islanders is smaller, less complex or more local than that of the New York City dweller but rather figure out the ways in which it de facto is, or is not, as well as "how this arises from, and has implications for, respective social organizations" (Barth 1978a: 11). Against this background Barth then urges us to follow up the potential of such an approach to scale, which focuses attention on the relationships between different levels of social activities, in terms of conceptual development.

Barth himself calls for an anthropological understanding of scale, which in his view would inevitably require researchers to rework his preliminary definition relying on size and population number as core elements. He even ends his introductory remarks by pointing out that "also the illusions of scale can have [real] consequences" (Barth 1978a: 12). This raises the question of how different people imagine and deploy scale when engaging in practical politics (cf. van Schendel 2005: 11). Barth, however, had difficulties to reconcile the idea of scale as a 'dimension of' social practice and political struggle with his vision that social life is 'occurring at' different scales (Mansfield 2005: 468). Nevertheless, many of the critical issues raised by him have independently been taken up in subsequent scholarly debates. While the body of literature is by now rather large (Herod 2011: chapter 1), it will suffice here to foreground an apparently unimposing shift in perspective from what has been called a 'politics of scale' to a 'scalar politics' (MacKinnon 2011).

The first term came up during the 1990s and quickly turned into one of *the* catchphrases among scholars seeking to explain how scales are made and unmade in struggles between social actors with conflicting interests and unequal access to various kinds of resources (Swyngedouw 1997a). Such analyses proceeded on the assumption that people are always already placed within a 'partitioned geography', which describes both the outcome of earlier rounds of scalar construction and the starting point for establishing new "geographical structures bounding political, economic and cultural activity in specific ways" (Smith 1995: 63). Scales are thus both the product and the progenitor of social processes (Smith 1993: 101), or, as Swyngedouw (1997b: 169) puts it, they portray "temporary stand-offs in a perpetual transformative [...] socio-spatial power struggle".

A major insight that can be drawn from this approach is the malleable character of scale, which has often been explained with reference to the ongoing movement of capital between locations (Brenner 1998). Various actors – from business elites to homeless people – appropriate, transform and subvert scale-dependent structural opportunities and pressures such as the vertical differentiation of labor and capital across dominant, or nodal, and marginal, or subaltern, scales (Smith 2004). More recently, however, this political economy perspective faced severe criticism, especially because of its alleged reliance on a flawed ontology (Marston et al. 2005;

Jones et al. 2007); the belief that a vertical hierarchy of scales – the largest is the globe, and the smallest is the body – *really exists* and that the global exercises some kind of causal power over 'lower' scales "which are bracketed with agency and practice" (MacKinnon 2011: 21; cf. Moore 2008: 212).

But instead of abandoning the concept of scale altogether, as some critiques demanded, scholars started to break new ground for discovering and conceptualizing how, as Latour (2005: 184) puts it, "[s]cale is the actor's own achievement". Remarkably, these endeavors put one of Barth's (1978b: 164) original concerns back on the agenda again, namely his call to scrutinize scales "in the real life of whole persons". The shift from a 'politics of scale', or a politics in which scale forms the subject of controversy, toward a 'scalar politics', or a politics in which actors deploy scale to achieve their objectives within (conflictual) social interactions, promises to fruitfully advance such a focus (MacKinnon 2011: 29). In the latter case it is not only the materiality but also – and perhaps even more importantly – the performativity of scale that gains significance (Kaiser and Nikiforova 2008).

In terms of performance scales mark less the external targets of people's activities and struggles, but are rather implicated as an integral dimension of what they actually do, believe and strive for. Based on a remark by Neveling and Wergin (2009: 333) one may therefore say that scale is "more than a framework – it is what makes the frame work". Scales assume significance within the strategies that actors deploy to advance their interests, such as when seeking recognition as a Scheduled Tribe. Whilst some actors might favor localization to effectively voice their concerns others might opt to regionalizing, nationalizing, or globalizing their strategies. However, the question as to what extent a contentious scalar politics is implicit within processes of ethnic identification has until now largely escaped consideration.

In addressing this shortcoming of contemporary research, I analyze how the Bhotiyas have leveraged the performativity of scale through various narratives and social practices to effectively voice their demands for tribal status in postcolonial India. The argument is divided into four empirical parts: As an entry point into a broader discussion of tribal identities and scalar politics, the first part introduces the narrative of Old Lady Jasuli, a widow who lived in the Darma valley during late colonial rule. The activities of the Kumaon Bhotiya Peoples' Federation (KBPF), which was founded shortly before India's independence in 1947, are analyzed in the second part. Its' members re-assembled representations that the British had coined about Kumaon's trans-Himalayan traders, especially in terms of gender relations. In so doing they articulated the Bhotiyas' tribalness by producing *Bhot* (or the region inhabited by the Bhotiyas) as a scale effect in independent India. This effect of scale materialized as a 'real' frame of social action when state officers notified the Bhotiyas as a Scheduled Tribe (ST) in 1967.

Only a few years later, as explored in the third part, a new generation of politically organized Bhotiyas challenged and transformed these inherited scalar categorizations and arrangements, which were then perceived to reinforce rather unequal power relations within the ST-community. Having organized themselves within various institutional frameworks, such as welfare and student organizations, these actors popularized the life story of Old Lady Jasuli. As a de-sexed widow who is dedicated to the common good, this local heroine embodies a powerful representational trope of Gandhian nationalism. It is shown how her character was deployed to convey an alternative sense of a tribal identity by reproducing independent India as a scale effect in *Bhot*. In the fourth part I then analyze how caste Hindus challenged the Bhotiyas' tribal status when Uttarakhand was carved out as an independent state of the Indian Union in 2000. These new competitors staked the claim that the Bhotiyas' ST-designation is a scalar expression that does not pertain to a specific people but to a region, including all its 'native' residents.

3.3 Old Lady Jasuli

On my very first visit to the upper Darma valley I travelled with the employees of a local trekking organization. Besides me, the guides escorted a small group of people from Kolkata who had come to visit what the advertisement poster promotes as 'the heights of virgin valley': a newly established base camp that is situated in a most picturesque landscape at the edge of Panchchuli glacier. Although the number of visitors booking the tour was still manageable, requests were rising steadily. A major reason for this positive trend is the fact that the 6-day trek – unlike those to many other high altitude destinations in the still rarely visited parts of northeastern Kumaon – requires no special permit.

The Lipu Lekh pass, which connects the neighboring Byans valley with the Tibetan plateau, was re-opened for an annually fixed number of Indian pilgrims in the early 1980s and for local traders in the early 1990s.[11] Dharchula Tehsil, however, remained enclosed within the so-called 'Inner Line' until the mid-1990s. This administrative arrangement imposed strict regulations on access to non-local people due to concerns of border security (Statesman 1981).[12] Local residents often complained that these measures inhibited economic development and the emergence of a regional tourism industry (Gerwin and Bergmann 2012: 100). The situation only started to change after Uttarakhand achieved its independence from Uttar Pradesh in the year 2000. It was thereafter, in 2003, that the trekking organization was founded as a joint venture with the public sector tourism enterprise called *Kumaon Mandal Vikas Nigam Ltd.* (KMVN).

[11] Between 2008 and 2010 the government issued approximately 40 trade permits per year. The main exported goods were agricultural products, hardware and cosmetics whereas wool, animals, garments and electrical items are imported. The number of Indian pilgrims who are allowed crossing the pass in order to visit Mount Kailash amounts to an average of about 600 persons per year.

[12] 'Tehsil' is an administrative unit in India that stands below the district level. The still contested borderline between Indian and China became officially known as the Line of Actual Control (LAC). The so-called Inner Line, which runs parallel to the LAC, "attempts to ensure that denizens of 'protected areas' remain pure and secure" (Aggarwal 2004: 57). Entry into Inner Line areas is strictly regulated and monitored by various government security agencies.

I made the acquaintance of the company's owner, Mr. Kutiyal, in a textile shop on the bazaar in Dharchula. Over a cup of tea we quickly got into a conversation. As a keen mountaineer and smart businessman Mr. Kutiyal was very excited about the gradual opening of his home region, because he had ambitious plans for furthering its touristic potential. The great advantage that he claimed to have over 'outside' competitors is a good personal network in the administrative circles of Uttarakhand, where tourism has also become a flagship project, as well as an excellent standing among local residents.[13] Community leaders strongly support the organization's endeavors, not least because they offer new income-generating opportunities. When I told Mr. Kutiyal about my proposed fieldwork in the upper Darma valley, he instantly offered me to temporarily join his team. In that manner I could both provide some feedback to their overall setup and establish contacts among local residents much more easily. And so I gratefully did. While this inaugural trip allowed me to register my presence and research intentions among Darma residents, the villagers also introduced me to a local heroine from colonial times. Her life story, they emphasized, had been nearly forgotten in the oblivion of history.

It was on the third day of our tour. I had split off from the trekking group and travelled along with a young men from Nagling village to some other nearby settlements. On the way we took rest on a small hill that offered a splendid view of Danthu village. As I was told, the position of that settlement within the valley is central not only geographically but also in terms of religious practice and belief. The seat of the valley's divine ruler, *Gabla Sai*, is located on a level ground slightly above the main village. This place, which is called Helipang Danthu, is frequently mentioned in the ritual chants of the Darmani as a meeting place of the valley's deities. Close to it one can easily recognize, even from a distance, the statue of a Bhotiya woman emptying a sack filled with coins and precious metals (Photo 3.1). The villagers address that woman as Old Lady Jasuli. My companion provided me with a rough summary of her biography, which I was later re-told in more detail by other persons:

> Old Lady Jasuli was the wife of a wealthy trader from Danthu village. Having one beautiful daughter and one handsome son, they could have led an easy and pleasant life. But Jasuli's husband died in a tragic accident on one of his yearly visits to the trade mart in Gyanima [located in western Tibet]. Her daughter married a respectable man from neighboring Dugtu

[13] In 2011 Uttarakhand was the only mountain state to be listed in India's list of top-ten domestic tourist arrivals (Business Standard 2011). 30,972,000 domestic and 136,000 foreign tourists visited the region in 2010 (UGP 2013). The young federal state advertises itself as *devabhumi*, 'the land of the gods' (Sax 2009: chapter 3). While the same term is also used in the neighboring areas of Himachal Pradesh and Nepal, Uttarakhand has perhaps the best claim, since it contains so many world famous pilgrimage centers: the so-called *chota char dham* or 'four abodes' of Badrinath, Kedernath, Gangotri and Jamnotri are located in Garhwal; at the foot of the Himalayas and on the banks of the holy Ganges river lie Rishikesh and Haridwar; and the main route to sacred Mount Kailash and Lake Manasarovar in western Tibet leads via Kumaon. That the federal Government of Uttarakhand has stepped up its efforts to particularly promote tourism is indicated by the fact that it was the first Indian state to create a separate Tourism Development Board in 2002, for which it received the prestigious National Tourism Award a year later. See UTDP (2008) for Uttarakhand's tourism development goals until 2022.

Photo 3.1 Old Lady Jasuli in Danthu village (Photograph © Christoph Bergmann, August 21, 2010)

village, but not even that could comfort her for long. Out of the blue and only a few weeks after the wedding ceremonies were over, the widow's son also passed away. Thus no male heir was left to administer the family's extensive property and great wealth. The widow, in her pain and deep sorrow, used a large amount of her wealth for the death rites.

She invited relatives and guests from near and far and hosted the entire village with food and drink for an entire year. But still her grief was not assuaged: it went on and on, seemingly forever! The roasted grains that are thrown on the floor while singing the funeral chants eventually built up to such a huge pile that the villagers were able to feed their livestock for several months from it. And Jasuli also prepared so much beer and spirits that these beverages formed an almost inexhaustible streamlet through the entire village, as though the jars in which the alcohol was stored had been spilt on the floor. And did the funeral rites finally come to an end? Yes, they did! And what happened then? Believe it or not, in all her sorrow Jasuli offered little pieces of gold into the Darma river. She did so day after day, hoping that these riches would reach the souls of her deceased husband and son. She put all of her hopes into these offerings.

Nonetheless, her grief grew greater with every passing day. Jasuli remained deeply unhappy and dissatisfied with her life until some *angrezi* [British] officer passed through Danthu village during an inspection tour. When he spoke to the village elders they told him about Jasuli, her great misfortune, and the fact that she was throwing her riches, piece by piece, into the river. The officer was moved by this story and came up with a striking idea: Jasuli should use her wealth to do something good, namely to furnish the major halting places of the region with proper rest houses. Not only would these houses make the journeys of the

> Darma traders more comfortable, but they would also be of great value to many others who might frequent those routes for various purposes. And not least, they could make Jasuli once again eager for life.
>
> When the British officer presented his idea, everybody, including Jasuli, got immediately excited. Cheered on by the crowd, Jasuli handed over her riches to the Englishman who kept his promise and guided the construction of these public facilities. All houses had at least four rooms, some even twelve, and none of them was equipped with a door in order to keep them open to all travelers. They were built along the foot trails connecting Taklakot and Gyanima in the north with Doti and Achham in the east as well as with Tanakpur in the south.

Although I am not able to confirm the course of these events through archival records, some of the rest houses attributed to Jasuli still exist in Kumaon. Regardless of the veracity or otherwise of the narrative, it stands out by explicitly referring to the time of late colonial rule. Previous research has created the impression that the most prominent oral traditions of the Bhotiyas would bracket the time of British colonial rule, i.e. a time when many Bhotiyas became wealthy and powerful (Brown 1984: 7; 29–34). As an example Brown mentions the legend of Sunpati Shauka, a wealthy trader and former ruler of the Johar valley, and his daughter Rajula. This legend forms an integral part of the popular Kumaoni folk culture. The story takes place at the time of Katyuri rule and thus before the eleventh century A.D. A short summary of it may help us to better understand why Old Lady Jasuli has become such an important character for the Darmani and why it is of significance that *her* narrative is set at the time of late colonial rule (Meissner 1985; Budhwar 2010: 91–109).

Longing for a child Sunpati Shauka and his wife conducted a pilgrimage to Haridwar, where they met a Tibetan couple of noble descent. Both families promised each other to intermarry their yet unborn offspring after a temple priest had foretold them that Sunpati would soon receive a daughter and his Tibetan counterpart a son. And so things went on. Sunpati's wife gave birth to Rajula, who became a most beautiful girl. As a young woman Rajula accompanied her father to Bairath to barter salt for grains.[14] While herding their flock at a nearby pasture, Rajula met Malushahi, a Katyuri king, with whom she fell in love. When Sunpati discovered the affair he became enraged, immediately took his daughter and hit the road toward Tibet. But Rajula escaped underway and fought her way back to Malushahi. After arrival at his palace she put him to a test of fidelity, which he accidentally failed and therewith made Rajula run away in disappointment. As a last resort Malushahi became a yogi and went in search of his beloved one. He found Rajula in her native village and started to secretly visit her every night in the guise of a parrot. But Rajula's parents got wind of what was happening and poisoned Malushahi. The story still found a happy end when a powerful *guru* from the royal court came to recover Malushahi with his magic. Through the *guru's* help the king was finally able to win Rajula for good and take her to his seat in Bairath.

[14] Bairath is considered to be the seat of Malushahi and according to the legend located in vicinity of Dwarahat, the former ruling center of the Katyauri dynasty in Kumaon. Dwarahat is located approximately 70 km northeast of Almora.

In its traditional format low-caste musicians would sing the story at weddings and at public fairs throughout the year. Nowadays people can also listen to Rajula-Malushahi-MP3s on their mobile phones or watch screen adaptations of the romantic affair on VCD and on YouTube. Such wide availability has enhanced the enormous popularity that the narrative had anyway already enjoyed. The plot can indeed be interpreted as emphasizing the old-standing association of the Johari with the (non-immigrant) Rajput-royals of the region, whom they gradually outpaced in terms of wealth and influence during the times of British rule (Brown 1984: 123–133). However, the story is worth looking at from another angle as well, one which links up with the actions of Old Lady Jasuli and concerns what Chatterjee (1989a) calls the 'resolution of the women's question' in postcolonial India. It was this question, or the control over women, that emerged as a key representational trope when the members of the KBPF articulated their demand for special statutory safeguards and constitutional protection as an ethnic minority. In the course of their negotiations they crafted a socio-spatial imaginary that accentuated the Bhotiyas' tribalness by producing *Bhot* (or the region inhabited by the Bhotiyas) as a scale effect in independent India.

3.4 Producing *Bhot* as a Scale Effect in Independent India

Rajula seems to fit perfectly within the dominant public image of tribal women as enjoying greater autonomy in decision making and a better status than high caste Indian women (Xaxa 2004). Jain (2009: 34), for instance, sees in her "an entirely new image of a woman – decisive, agile, full of esteem and self-empowered, and above all, ready to confront the biggest of her foes", viz. patriarchy. Such images are not only highly romanticized but also quite popular. In a widely sold Indian newsmagazine, which published a community portrait on the Bhotiyas in the mid-1970s, the front-page photo carried a corresponding caption: "No fear for women's lib. The Bhotia [sic] women live on equal terms with the menfolk and are allowed to choose the man they want to marry. They do the same work as the men and compete with them in every field" (IWI 1974: 30). There are at least two major driving forces that contributed to the popularity and pervasiveness of such images. On the one hand there is a grain of truth in them: tribal (and low caste) women tend to have more freedoms as their high caste counterparts in *purdah*, especially because they contribute more to the domestic economy.[15] On the other hand colonial administrators all over South Asia captured and popularized narratives of gender equality and sex-

[15] It is worth mentioning in this regard that Bhotiya women took a lead in the so-called *chipko andolan* during the 1970s. The *chipko andolan* is an environmental movement that started in Garhwal. The name *chipko*, which means 'to adhere', refers to the way in which people hugged trees in order to prevent them being felled by forest contractors (Rangan 2004; Linkenbach 2007). After the independence of Uttarakhand Bhotiya women often served as a role model for female empowerment in the state (Klenk 2004).

ual liberality among tribal groups, which then infused public perceptions as well as tribal policies and studies (Xaxa 2004: 355).

This latter observation also applies to the case of the Kumaoni Bhotiyas. Already Traill (1992 [1832]: 118) annotated that among these people, "[i]n the institution of marriage, the inclinations and will of the female appear to have greater weight than is common in the East, both with regard to the formation of such engagements and in the subsequent domestic management". Later observers then progressively registered how gender roles transformed in the different Bhotiya valleys. Sherring (1974 [1916]: 68), for instance, remarked that

> [i]n Johar, with the pride of new converts to Hinduism, the men in every family eat first, and then the women, the leavings being always for the women and the children, whereas in Darma pargana [an administrative unit referring to the three valleys of Darma, Byans and Chaudans] there is still found the old-fashioned politeness, which shows some regard for the feelings of the gentler sex, and which acknowledges no custom of eating leavings, for in all families men, women, and children all sit down and eat together. Further, throughout all the domestic details of life the eastern Bhotias invariably put women in a high place, in this way differing radically from Hindus, who regard her as a chattel, and Buddhists, who let her rank among dogs.

By describing gender roles, in other words, British obervers could not only distinguish *Bhot*, or the region inhabited by the Bhotiyas, from surrounding areas but also highlight the divergent developments that occurred within *Bhot*. These developments were conceptualized in terms of Hinduization, i.e. the acculturation of a group to Hinduism. Among the Darmani, who were always considered to be least influenced by outside forces, women would still enjoy "the greatest liberty" and feast on "the romance and poetry of life" with their male compatriots (Sherring 1974 [1916]: 72). But it was not only the 'lax' regulation of domestic affairs, but also and especially a communal institution called *rambang* that attracted the attention of and stimulated discussions amongst outside observers. *Rambang* was held in village clubhouses where unmarried teenage boys and girls spent the nights singing, dancing and drinking. The Johari claimed to have abandoned this institution long before colonial interference. But it was not only them who came to look down upon those who maintained it. Torn between fascination and deprecation Sherring (1974 [1916]: 104) also described it as a rather disreputable place:

> Girls start to go to the Rambang from the age of ten years, and practically never sleep at home after that age, the result being that a virtuous girl is scarcely known in the pargana [administrative unit] of Darma. As is to be expected, a system such as this leads to the freest intimacy, and one sees a man walking about with his arm round a girl's waist both under the same covering shawl, a practice common in Europe but rare in the East: modesty is unknown and there is a boldness in the faces of women. Intentional miscarriages of illegitimate children are not at all uncommon.

Already B.W. Colvin, an interim district officer in Almora during the Indian Mutiny in 1857, had agitated against *rambang* for its assumed negative social impact (Pant 1935: 228, footnote 2). At that time, however, the British remained virtually absent in the region inhabited by the Bhotiyas so that his endeavor amounted to nothing more than lip service to colonial engagement on the northern frontier of the empire. The debate first received a major impetus when missionaries from the U.S.-based Methodist Episcopal Church settled in Dharchula in the early 1880s. They closely attended to the activities and developments in *pargana* Darma, and regularly exchanged their views with British administrators. One of the mission's founding members, Ms. Browne (n.d.: 22), wrote on the same subject that

> [c]ontrary to the custom in India, there is free intercourse between the sexes in Bhot. [...] Young people are perfectly free, and, as far as intercourse with the opposite sex goes, parents have no control over their daughters and no voice in marriage arrangements. If a young man has serious intentions towards a young woman, the parents of either party must on no account be informed. Each village has two or three Ramang [sic] houses, i.e., clubs belonging to the girls of the village, where they spend their nights together, instead of sleeping in their own homes. To these houses resort the young, or even married, men, generally of other villages. The night is spent in spinning, singing, drinking and laughing and joking till overcome by sleep. The evils of this custom are beyond estimation, and one will readily believe there is very little virtue in Bhot.

Of course, the missionaries undertook many efforts to disband these institutions and to teach young people, especially women, a 'better' sense of decency and morality. They tried, for instance to substitute the 'suggestive' chants sung in *rambang* by vernacularized versions of the gospels. However, results remained, according to the missionarys' own statements, mediocre (Browne n.d.: 61). Still, these discourses and representations had political repercussions. While egalitarian gender relations were considered by outside observers as a common denominator to distinguish 'the Bhotiyas' from 'the Hindu mainstream', the customary institutions that contributed to their maintenance, such as *rambang*, came to be devalorized by an increasing section of the community itself. This devalorization happened, however, for reasons that were slightly different from those of both British administrators and North-American missionaries. In order to track this difference, a brief look into the most comprehensive pre-independence ethnography on the Kumaon Himalaya is required. The work, which was written by Pant (1935: 11), spells out that

> [t]he structure and character of the Himalayan's home has also been largely determined by the physical environment. Thus, while among the Bhotiyas habits of nomadism have led to the development of a sturdy and independent womenhood, which produces sensible wives and competent mothers, among the hillmen, on the contrary, a narrow and bigoted conservatism has degraded women to the position of a miserable drudge. Nor should we omit to notice that striking contrast between the full and free participation of the Bhotiya woman on equal terms with their menfolk, in work and play, and in social intercourse in general, and the total exclusion of the Sub-Himalayan woman from a share in all those activities that lend a variety and charm to life and make it worth living and striving for.

The book was originally published as an M.A.-thesis that Pant, himself a Kumaoni, wrote from the University of Lucknow under the supervision of Radhakamal Mukerjee, a founding member of the so-called 'Lucknow School' (Sundar et al. 2000). That school was one of *the* intellectual hubs within the Indian national movement, where Mukerjee was well known for his critical stance towards the idea that Westernization describes the only possible way forward. In terms of a 'social ecology' he underlined that communal institutions, as well as the customary practices and norms through which they become reinforced, play a crucial role in keeping regional communities adaptive and vital. Instead of substituting these eastern institutions by western ones, Mukerjee promoted the idea "that India can fruitfully preserve and tap her 'communalistic' or community-oriented institutions for evolving a path of progress more suited to Indian conditions" (Joshi 1986: 1460).

Pant (1935: 11) followed his teacher in important respects: ecological constraints in northern Kumaon favored the choice of a mobile-pastoral existence, which in turn set the course "for a sturdy and independent womenhood". According to him Bhotiya women act as the bearers of societal values that are representative of the whole community. It is noticeable, though, that Pant borrowed these values from the European enlightenment, viz. freedom and equality, and then tied them to Indianized domestic role-models: the 'sensible wife' and 'competent mother'. At least since the 1920s, when the Gandhian nationalist movement gradually turned into a mass phenomenon, motherhood had evolved as a key trope for distinguishing Indian from Western women (Hancock 2001: 896; cf. Chatterjee 1989a, b). And it was within this emancipatory framework that Pant reasoned about and explained the relevance of an institution like *rambang*, which might at first sight appear rather strange when compared to the middle-class patriarchial norms of mainstream nationalist ideologies (Pant 1935: 228):

> The women-folk of these nomads, especially those of *pargana* Darma, are left alone by the men for a considerable part of the year. This has led to the development of clubs or guilds, composed entirely of women or girls of a particular age. The young folk, whenever present in the villages, are privileged to be admitted freely to these clubs, but the management rests primarily with the unmarried girls and such of the married women as are not yet mothers.

Although Pant does mention that these institutions permit premarital sexual experiences, he differs in one crucial aspect from all previous accounts by colonial administrators and missionaries: throughout the book, there is not one negative remark about them. *Rambang* rather served to illustrate how young women grow into their designated roles as wives, mothers and, perhaps, also as lovers. In that institution women would develop their managerial capabilities, autonomy in decision making and feminine charms. But about a decade after Pant's book had been published, the KBPF presented its own view of the situation and declared *rambang* as a prime target for social and educational reform. In their report from 1947 the institution is discussed under the unmistakable heading 'The gloomy prospect':

> Venereal diseases have spread among them [the Bhotiyas] on account of two social conditions. First a Bhotiya remains for 10 months of the year away from his family on his trade and business. This puts a great strain on his moral stamina, which mostly fails him. This is true of all the Bhotiyas of Kumaon but another social practice, which is prevalent only among Pargana Darma Bhotiyas, is also playing havoc with their social life. This is the institution of Rang Bang [sic] which is prevalent only among patti [valley] Darma Bhotiyas at present. With the loosening of old taboos and other methods of social control this institution though common in all the civilized countries of the West and which once fulfilled certain social needs, seems to have out-lived its usefulness in the present day society of the Darma, Bians [sic] and Chaudans Bhotiyas.
>
> [...] Under these circumstances, the spread of education among the Bhotiyas of Kumaon is the only hope through which the evils of ignorance can be banished and people made to realize their duty not only towards themselves but also towards their society as a whole. At

> this stage, it is the duty of the Govt. and public men that the evil practices of the Bhotiyas are abandoned through education and if necessary through legislation. The certain customs as to marriage widows and divorce etc. laid down in Dr. Panna Lal's [sic] 'Hindu customary law in Kumaon' relating to the Bhotiyas of Darma, Bians and Chaudans are proving mischievous and anti-social and need immediate suitable amendments, taking into consideration their ultimate social benefit. This is on account of the fact that certain local customs have outlived their utility.
>
> With due consideration of these evil facts which is eating up the very vitals of society, plans shall have to be made for the mental, moral and cultural uplift of the Bhotiya people. (KBPF 1947: 15–16)

The Federation adopted the anthropologists' language of functional utility to rationalize what was in fact an attempt to reform their society so as to make it more acceptable to mainstream Hindus. The need for reform was necessitated by the emergent political setup of independent India and the decreased interest of the Bhotiya elites in following the old trade patterns. In light of these transformations *rambang*, the 'traditional' institution of female action and authority, was considered outdated and in need of replacement. Education and – if necessary – legislative measures should advance a moral transformation in order to make 'the Bhotiyas' fit into an incipient national community. This amounts to the following picture: an ethnic group tries to raise its overall status within a larger, multi-ethnic society at the expense of its women, i.e. by imposing restrictions on female community members.

The existence of clubhouses has long fuelled outsiders' imaginations about the sexual permissiveness amongst India's tribes (Xaxa 1999: 1520). In Kumaon, scholars also tried to proof such imaginations 'scientifically'. Tiwari (1952: 150, footnote 6), for instance, who conducted research on the distribution of blood groups among the Bhotiyas, pointed out that "Rambang makes sufficient allowance for intermixture with their neighbouring people". As in other parts of India, the tribal dormitories would allow for free intimacy and sexual relationships even to the foreign visitor. Srivastava (1953: 202, footnote 2) harshly criticized this by saying that "Mr. Tewari [sic], who only remained at Dharchula, would have held a different opinion if he had toured the whole region and observed rangbang [sic] gatherings in action himself". While similar appraisals have been widely reported across South Asia (Stoler 1997), the developments resulting thereof took a specific trajectory amongst the Bhotiyas.

Rambang served the KBPF to indicate the uneven success of the different valley communities in keeping pace not only with the process of Hinduization – as it was before – but also with the formation of a modern Indian nation. Hinduization, or Sanskritization, and Modernization, or Westernization, are therefore perhaps not as opposed as Srinivas (1956) has originally proposed in his classical essay on the topic.[16] It was as modernist and reformist citizens that the members of the KBPF affirmed their right to negotiate with, be recognized by, and derive benefits from the state. However, by asserting their right to Scheduled Tribe status the Federation had to simultaneously relate certain cultural distinctions to a geographically confined

[16] The term 'sanskritization' describes the process by which caste groups are trying to raise their status by emulating the practices of the highest castes.

people whom the officialdom at the higher echelons of the Indian government could then easily recognize as tribal, i.e. as 'primitive' in terms of Hinduization and 'non-modern' in terms of societal values.

For those sections of the community who still practiced *rambang* the situation appeared ambiguous. On the one hand they were seen to compromise the honor, reputation and esteem of the Federation's aspiring members. On the other hand it were their 'primitive' customs that legitimized the Bhotiyas' entitlement for special statutory safeguards and constitutional protection as a national minority. In any case, the KBPF regenerated a by then already time honored status hierarchy. Those who had already abandoned that institution and therewith discarded the 'evils' of the past would have taken a first step towards realizing their individual and societal duties as national citizens. All others, namely the residents of Byans, Chaudans and Darma, would still have to achieve this. *Rambang* (which literally translates as a 'place to come to') became even referred to as *rangbang*, or 'the place of the Rang', with 'Rang' being the locally used ethnonym of the eastern Bhotiyas. And among them, the Darmani stood out as the most inferior group since only they openly kept that 'outlived' institution.

But even after India's independence the issue remained subject of diverging views both within and outside the community. When Srivastava (1953: 193) visited the Darma valley for ethnographic fieldwork during the early 1950s, he experienced *rambang* as nothing scandalous but as an inoffensive "entertainment and a recreation after a hard day's work". It was presumably the gaze of predominantly male outsiders who had emphasized an erotic aspect, which was 'traditionally' not so prominent. When I interviewed villagers in 2009 some did not like talking about it, others had a negative view, while still others described it as a playful event that involved the amusing display of talent (singing and dancing), stamina (drinking), and cunningness (impressing others, in particular unmarried girls). While participation was regulated according to exogamous clan affiliations, *rambang* was neither the only nor the most prevalent procedure for arranging marriages.[17]

In the Darma valley it was very common that families promised each other an intermarriage between their children when these were still at a very young age or sometimes even unborn (Leder 2003: 91). Cross-cousin marriage was a widely followed marriage pattern. If a couple, for whom such arrangements had not been made, developed affection in *rambang*, the boy would have eventually proceeded by capturing the girl and taking her to his native village for making an official marriage proposal. While the girl always had the right to reject, an unmarried cross-cousin could also, and as a question of honor, re-capture the girl within a certain time period. But all these customary practices faced increasing outside pressures for change. Srivastava (1953: 193) summarized the situation as follows:

[17] Lall (1942: 29), to whom the Federation referred in the aforementioned quote, even maintained that while *Rambang* was the most crucial platform for arranging marriages in the Byans and in the Chaudans valley, it had no such function at all in the Darma valley. The Darmani would simply maintain these gatherings out of 'infamy' and 'corruption', and they would be eager to hide their activities so that no court could ever blame them.

> Bhotias who came to Almora for education, went back preaching to their people the evils (!) of their age-old customs and practices. Coupled with these enthusiastic reformers who had suffered humiliations among the Hindus at Almora and were convinced of the evil of all their institutions, dissimilar to those of the Hindus, the influence of Hindu pilgrims passing to Kailash and the ascetics who have established their *ashramas* in these regions, has been no less. Young boys and girls, dance and sing collectively. They do, of course, now take particular care to conceal every information about these meetings, and tell openly that they do not have the Rangbang any more.

This quotation highlights some of the key dynamics that intensified the discussions about *rambang* and contributed to its bad reputation: In 1936, Narayan Swami had established an *ashram* and two schools in the Chaudans valley near Pangu village. In that village the defamation of this institution took extreme forms. Residents went to enormous lengths to prevent youngsters engaging in it and also crusaded against the widespread consumption of liquor (Das and Raha 1981: 259). Whilst the agitation achieved only moderate success in other villages, it heated up the debate among and between the residents of different valleys. In the Byans valley, the former outpost of the *political peshkar* in Garbyang turned into another important bastion of the reformist movement. Residents of that village hold important positions in the KBPF, and their demands steered especially in the direction of improving educational opportunities for the young.[18]

After the war between India and China in 1962, the KBPF and its local offshoots invested even more efforts in obtaining official ST-recognition. When they finally succeeded in 1967, this guaranteed them quota-access to universities, government services and the legislature, amongst other privileges. The central government had supported these endeavors in opposition to the then chief minister of Uttar Pradesh and the majority of the Pahari population (Husain 1995: 11). One rationale given for the recognition by government officials was the potential drift of the Bhotiyas towards China if not supported on the Indian side (Gerwin and Bergmann 2012: 100). By opting for ST-status, the Bhotiya leaders had intentionally positioned themselves at the margins of Kuamon's caste society. At the same time they promoted discourses and representations that facilitated their access to the class-ridden spheres of power and wealth across various administrative levels. In this context employment in the administrative centers of Uttar Pradesh became feasible opportunities for the Bhotiyas' educated elite.

Following the awarding of Scheduled Tribe status also the Darmani rhetorically and practically distanced themselves from *rambang*. In effect, they changed its

[18] A highly influential female personality from that village, Gangotri Garbyal, efficiently represented this agenda in public (Leder 2003: 131–132). Born in 1918, Miss Garbyal was the first Rang woman to obtain a university degree. When her fiancée died before their wedding day, she publicly announced herself a widow with the intention of devoting her life to the education of children. After several posts in intermediate colleges she was appointed as the first female vice president of the supervisory school authority of Almora District between 1948 and 1952. Thereafter, she became actively involved in several social work projects in the valleys of the eastern Bhotiyas, many in cooperation with the Narayan Ashram. She strongly supported the Darmani in advancing reforms and also played a role in popularizing Old Lady Jasuli, who was, like herself, a widow dedicated to the service of those who suffer (Dhakriyal 2004: 300).

name once again in everyday parlance. Whereas the KBPF had referred to this institution as *rangbang*, so did people then, with a pinch of irony, identify it as *Doti rambang*. Doti is a village in far-western Nepal that marked an important transit point for the migrating Rang. The village fell into disrepute – especially among Rang women – for ostensibly providing opportunities for men to have extra-marital affairs.[19] While *rambang* was thus finally expelled across an international border, gender roles, as is shown in the following part, remained an important representational trope. They were again revived by Bhotiyas who sought to re-envision their tribal identity within the webs of relations that were opening up to them as members of a Scheduled Tribe.

We have so far seen how members of the KBPF skillfully assembled representations that the British had coined about the Bhotiyas in order to articulate their demand for special statutory safeguards and constitutional protection as an ethnic minority. Colonial officers had described both the cultural distinctiveness of 'the Bhotiyas' vis-à-vis neighboring groups and the community's internal diversity in terms of gender roles, especially with regard to the position of women. The KBPF followed up on that by taking the enduring presence of a local institution, namely of *rambang*, as a source of legitimacy for their cause. Within that institution unmarried women could interact rather freely with men, a circumstance that the Federation considered as 'anti-social'. Coupled with the perceived lack of education, the Bhotiyas appeared as being hardly able to advance endogenous development.

Within these strategic efforts of the KBPF one can clearly see the performativity of scale at work. The Federation articulated the Bhotiyas tribalness by deploying discursive devices that produced *Bhot* as a scale effect in independent India. This effect of scale accrued from the socio-spatial imaginary of a remote mountain region whose inhabitants, the Bhotiyas, were afflicted by the 'evils of ignorance'. With the awarding of Scheduled Tribe status in 1967 this scale effect materialized as a 'real' frame of social action, as all Bhotiyas were then entitled to benefit from India's tribal policy. However, the scalar categorizations and arrangements that had been promoted by the Federation also reinforced rather unequal power relations within the ST community. A new generation of politically organized Bhotiyas then gradually started to re-envision the ties between the 'lower' scale of their tribal society and the broader currents of societal change in India. Particularly the Darmani, who had until then appeared as the most disempowered section of the community, should play a central role in this movement. With the aim of conveying an alternative sense of identity, they popularized Old Lady Jasuli's life and therwith reproduced independent India as a scale effect in *Bhot*.

[19] In the bazaars of Chamoli District in Garhwal, however, Nepali laborers are called 'Dotiyals', and widely respected for their very careful adherence to rules of pollution and contact. It is also often remarked how chaste their wives (who often accompany them) are (Sax, personal communication).

3.5 Reproducing Independent India as a Scale Effect in *Bhot*

Jasuli had to manage with a run of very bad luck. She not only became a widow but also lost her son. In a patrilineal society this typically amounts to a rather vulnerable position (Agarwal 1998). But unlike in many other parts of India a widow among the eastern Bhotiyas remains in possession of her deceased husband's riches – as well as debts – until she dies or asks her in-laws for relinquishment (Srivastava 1957: 41).[20] Nevertheless, Jasuli finds herself in a state of deprivation. This has nothing to do with a lack of economic assets, mistreatment by some greedy in-laws or any other social restriction and insecurity. She is deprived, rather, of knowledge, namely the knowledge and farsightedness for making meaningful use of her affluence. Jasuli does not struggle against outside interference, like Rajula, but rather takes pains to comply with emergent orders for making her husband's inheritance, to speak with Gandhi (1971: 74), "live again in her actions". And she strives towards this end not for personal pleasure but for the benefit and welfare of society.

Why do I refer to Gandhi at this point? In the 1920s Gandhi (1966: 123) had remarked that the "spinning wheel is the symbol of the chastity of the womanhood of India. In the absence of the spinning wheel I give you my testimony that thousands of our poor sisters are giving themselves to a life of shame and degradation". The spinning of *khadi* (a cloth from cotton, silk or wool) was turned into a nationalist project in which women, as the bearers of superior moral values, were considered indispensable. Through their maternal courage and intuition they were supposed to weave the outcomes of constitutional reform and collective public action into daily practices by nourishing, so to speak, the new classes of nationalized subjects right from the cradle. On the eve of India's independence the KBPF signaled that this work of *khadi* was a successful one in *Bhot*: "While the men spin, the women weave. […] Nowhere else have the industries attained such stability and become so inextricably interwoven with the whole texture of domestic and social life" (KBPF 1947: 8). The Federation had invoked the normalizing powers of education to exorcize the gloomy prospect, 'the life of shame and degradation', which institutions like *rambang* would propagate.[21]

Old Lady Jasuli, however, is not prey to any of those sexual desires, promiscuous consumption patterns and evil virtues that the Federation had uncovered. She rather echoes another aspect of female praxis that Gandhi came to hold in high esteem

[20] In Darma a widow is principally allowed to remarry, and in these cases she would be obliged to ask her in-laws for relinquishment. The deceased husband's family indicates their acceptance by giving the widow a white piece of cloth along with a small symbolic amount of money (Srivastava 1957: 41). If she starts a relationship without doing so, her new partner and any children would be considered illegitimate. In that case her in-laws would also be allowed to claim a substantial amount of money from her second husband. The issue of widow-remarriage was as well a target of reform by the KBPF, as the statement on *Rambang* quoted before indicates.

[21] That members of the KBPF were familiar with the Gandhian nationalist movement is likely. Gandhi himself toured through the Uttarakhand Himalayas in 1929. During that tour he delivered a public speech in Almora on the importance of education (Joshi 2001: 3306).

after a large number of women left the hearth and performed *satyagraha* (civil disobedience) in the streets during the 1930s. This was the

> received image of a Hindu widow living on the periphery of the society. [...] On one hand is the construction of a morally superior woman, a loving wife and a dedicated mother, involved actively in the making-of-the-nation, spinning *khadi*, sharing equal tasks in the house with the male, and on the other a reconstruction of a de-sexed woman, who has 'sacrificed' her reproductive aspects, her family life and has dedicated herself to the service of the nation. (Patel 1988: 379)

Jasuli thus corresponds to a woman in front of whom – to paraphrase Gandhi (1969: 443) – one should instinctively bow his head in reverence, because a *satihood* like hers is one of the prime gifts that Hinduism has on offer for humanity.[22] However, the popularization of her narrative describes not just an ideologically motivated preoccupation with a key trope of India's nationalist discourse but also a strategic effort to re-envision the ties between the 'lower' scale of a remote mountain region and the broader currents of societal change in India. To illustrate this point let us briefly go back to where Jasuli's narrative sets off in time, namely the late nineteenth century. It was back then that the upper valleys of the eastern Bhotiyas were, for the first time under British rule, more firmly integrated into the administrative apparatus of the colonial state. British administrators had long reckoned on the Bhotiyas' assistance for handling affairs on the imperial frontier. Jasuli, by contrast, entered into an alliance with a colonial officer to take over responsibility for the public good. But it goes even further. As local residents now see it, Jasuli required the officer's patronage not so much because of her personal fate as a widow, but rather for her (and the community's) want of literacy – of formal education – that neither she nor any other local resident possessed at that time. Jasuli's state-partnership not only resulted in the establishment of public goods away from the currents of daily village life, namely a number of rest-houses along the pathways in the middle ranges of the Kumaon Himalaya, but is also associated with a major change that occurred in the Darma valley itself: the opening of the first governmental primary school.

This tallies well with official historical sources. While the British had already maintained schools in upper Johar (Milam) and Byans (Garbyang) since the 1850s, the first school in Darma was not established until the late 1890s. Besides a privately run missionary school in Bon village, Walton (1911: appendix, xxiii) lists only one state-sponsored single-teacher primary school in Dugtu village, which together served 54 regular pupils for the year 1911. In comparison, there were already eight primary schools with a total of 245 pupils in Johar and three schools with a total of 131 pupils in Byans.[23] This provides a further reason why the Darmani played no

[22] The term *satihood* is commonly understood as indicating the perfection of an Indianized womanhood, with *sati* referring to a woman who dies on the funeral pyre of her husband. For Gandhi, however, *satihood* should not be accomplished through an act of destructive subordination but rather through deeds of sacrifice, dedication and compassion for the common good and therewith for the nation (Patel 1988: 384).

[23] In the uppermost settlement of Johar, Milam, Kumaon's first girls' school had opened its doors in 1867 (Walton 1911: 155).

formative role and held no strong position in upcoming political institutions like the KBPF.[24] Dominant leaders were all sourced from a well-educated stratum, a 'creamy layer', whose members had entered formalized career paths in the public sector long before India's independence.

Many people in Darma told me that even after India's independence, parents did not put much pressure on their children to go for further education, not least because some basic skills in accountancy and written Hindi sufficed to ensure them good commercial dealings and therewith a comparatively decent standard of living. Especially due to their geographical vicinity to and solid personal contacts in Nepal, the eastern Bhotiyas remained in control of alternative and economically sound pastoral routes, whilst trade volumes across the high passes into Tibet gradually declined. Far-western Nepal remained quite undeveloped in infrastructure until the late 1980s, so that large flocks provided a perpetual source of income as transport animals for construction materials, groceries, garments and Tibetan wool.[25] The relative lack of arable land in the winter settlements around Dharchula also kept the Darmani from casually abandoning their migratory lifestyle.

The Sino-Indian border war marked yet a decisive turning point. This is not so much because the international conflict brought tremendous suffering to resident populations – the region always remained a rather quiet sideshow of the war – but rather because young people then opted for lifestyles that differed radically from those of their parents. Pastoralism was still a lucrative undertaking, but increasingly considered as not being worth striving for any more. Children could suddenly observe the moving-in of alternative life opportunities right from their doorsteps, especially because the Indian state implemented a large military road-building program, which opened up the lower parts of all the Bhotiya valleys in Kumaon by the 1970s (Rawat and Sharma 1997).[26] Near Dharchula a huge military base was established that has remained a dominant element of the townscape up until now. A resident of Sipu (Sipal n.d.: 16–17; my translation from the Hindi language) village, who was a young boy at that time, remembers the changing scenery in Dharchula as follows:

> The everyday activities of the military became a subject of curiosity amongst children: when seeing their morning drill and day's parade, their war exercises and radio operations around the camp, their many types of training in the boxing ring, where two young men in strange costumes would engage in hand-to-hand fighting, our hearts would miss a beat. Sometimes we were delighted, at others scared. In Motiyadhar [a place near Galati, the

---

[24] There was only one person from the Darma valley who was listed in the KBPF-report from 1947 as a regular member.

[25] For more than a decade after the closure of the Sino-Indian border in 1962, the eastern Bhotiyas continued to trade Tibetan wool. They imported this commodity via Nepal, namely via the two Bhotiya villages of Tinkar and Changru. As people report, the profit rates were a good deal worse than before, especially since Nepalese merchants used their monopoly over the trade route towards Tibet for their own benefit. At that time a major destination for Tibetan wool in India was a wool mill in Panipat, Haryana.

[26] According to local informants the road from Askot to Dharchula was finished in 1962, from Dharchula to Thawagat in 1965 and from Thawagat to Sobla in 1982.

> winter settlement of the people from Sipu] there was a check post of the C.M.P (military police), where car drivers would always blow the horn horn due to the narrow turning of the road. Whilst the people from Sipu and Marcha would have always heard this sound, the subject of my curiosity were the children who returned from the Dharchula G.I.C. [Government Inter College] in the evening [...]. The pupils from Kalika, Gothi and Galati, when returning from Dharchula, would be rolling their self-made stick-wheels from iron wire. Meanwhile, I was impressed by the sound of school utensils rattling in their bags and the ink-marks on their fingertips and in their pockets.

Militarization appears here as a kind of backdrop for articulating a marked transformation in people's daily life-rhythms and aspirations, particularly amongst the young. Similar shifts were experienced in the upper Darma valley. There, the border crisis brought forth an even tighter entanglement of military and civilian lives. The Indian army heavily relied on the assistance of village dwellers, especially for logistical support, housing facilities and for espionage. The Darmani benefited from these emergent constellations not only financially but also in numerous other ways.[27] Even today people speak about the fact that the army stationed both a medical and a veterinary doctor in Dugtu village during the post-war years. Moreover, the District Board Pithoragarh made available more generous funds for improving the educational prospects in all the valleys inhabited by the Bhotiyas.[28] Also the so-called Darma-Khampas, who were until then hardly recognized by administrative bodies, received an elementary teacher who accompanied these Buddhist nomads year-round on their extensive migratory route between the Indo-Gangetic plain and the High Himalayas.[29]

The building of schools and the implementation of a decentralized administrative structure (Nautiyal et al. 2003) facilitated the gradual transformation of Dharchula into a densely populated administrative and educational center. After the Bhotiyas had officially been recognized as a Scheduled Tribe, the opportunities to acquire higher education improved greatly. It is still widely remembered when the first batch of pupils from the Darma valley left their homes amidst great rejoicing and tears in their mother's eyes to attend boarding schools outside of Pithoragarh District in the early 1970s.[30] In the following years a number of student associations were established to promote the benefits of education in the upper villages. The refrain of a song that was popularized during that time phrases the agenda as plain as it can be:

[27] Besides the availability of kerosene, villagers also began to directly sell parts of their cultivated field crops and hay to the military posts in the upper valley. The prices received are till the present day considered being good, not least since the transportation costs are comparatively low.

[28] Pithoragarh was carved out as a separate district from Almora in 1960–1961 for setting up an administrative structure that allowed for a better handling of the sensitive border region.

[29] The former summer camp of the Darma-Khampas, a place called Khimling (also referred to as Bidang) became an important gateway for military operations on the contested border. Till the present it is the last outpost of the Indian army in Darma. After the Khampas stopped migrating more than three decades before, some of them pursued an impressive carreer in the Indian civil and military service.

[30] According to the statements of local villagers, one of the first boarding schools visited by pupils from the Darma valley was located in Lakhimpur Kehri, Uttar Pradesh.

"Oh mother and father, instead of caring for our education why do you force us to graze the sheep?" (Sipal n.d.: 18; my translation from the Hindi language).

The Border Youth Welfare Association (BYWA) of Pithoragarh emerged as an important umbrella organization for educated community members from all three valleys of the eastern Bhotiyas. The association coordinated activities at the grass-roots level and supported poor families to receive access to educational institutions and thereafter employment for their children. This group also entered into an active dialogue with village elders and the established political elites who set the tone in other institutions. Noteworthy are the Darma Service Committee (DSC), which was already founded in the 1920s to regulate the commercial affairs in *pargana* Darma; the Rang Cultural Council Pangu (RCCP) that was formed in response to the activities of Swami Narayan during the 1940s; and the Byans Scheduled Tribe Uplift Committee (BSTUC) that was set up after the Bhotiyas had officially been awarded the Scheduled Tribe status in the late 1960s.

It was this institutionally organized post-war generation, the first to experience the benefits of the state's reservation system, which popularized Jasuli's narrative. The local historian D.S. Dhakriyal, who was one of the prime movers in the BYWA back then, wrote a remarkable statement in that regard. It can be read as a commentary on the discourses and representations that the KBPF had promoted in its report, which held that the eastern Bhotiyas, and especially the Darmani, would need to realize their 'duty' both towards themselves and the society as a whole:

> I am greatly surprised that this woman [Old Lady Jasuli], whose important achievements were a matter of public concern in both social and religious terms, was ignored in the articles published by Kumaoni historians and British officials. If a woman of some other caste or class had done something similar, then they would definitely have been properly acknowledged on the pages of history as well as verbally praised. But the actions of a woman who belonged to this ethnic group [the Rang] and lived in this corner of the Himalayas [the Darma valley] were not considered important enough and were therefore never publicized. These are the most probable reasons why her benefactions never did or could appear on the surface of public interest. [...] Her remarkable achievements and personality were rendered invisible through cut and dried opinions of casteism, classism and regionalism. (Dhakriyal 2004: 295–296, my translation from the Hindi language)

The major objectives of the BYWA were to combat this ignorance and to challenge the negative publicity that their community had hitherto experienced. During the 1980s the organization published a series of articles about Old Lady Jasuli in regional magazines. The first of them, entitled 'The Maternal Flavor of My Darma valley' appeared in 1980 in *Parvat Piyush* (*Mountain Nectar*), an influential conservative weekly released from Pithoragarh (Dhakriyal 2004: 300). As a de-sexed woman who dedicates herself in a motherly fashion to the common good, Jasuli's life history served to uphold a sort of civic responsibility that thrived by means of a special relationship to the state. Just as the British colonial officer had held his protecting hand over the desperate widow, so had the Government of India recognized the tribalness of the Bhotiyas. Through this representational practice *Bhot* did not appear anymore as a breeding ground for 'evil' practices and 'anti-social' customs but rather crystallized as an epitome of Gandhian nationalism.

This overall discourse was supported by a number of monographs written by community members. These publications delivered new interpretations of the historical, ethnic, cultural, religious and social distinctions that would qualify 'the Bhotiyas' as a tribe or *janjati*. The most influential were Raypa (1974), Garbyal (1987) and Pangtey (1992). All three authors have in common impressive careers in the public sector. It is further striking that they all adopted the ethnonym 'Shauka' for the people about whom they wrote. 'Shauka' is the preferred self-designation of the Johari and frequently used by all Kumaoni Bhotiyas when speaking and writing about their community in Hindi (Nawa 2000). And it was to a Hindi-speaking readership that these books were basically addressed, especially since they intended to disprove the dominant public perception of 'the Bhotiyas' as some sort of 'crypto-Buddhists' (Bergmann et al. 2008).

New hypotheses were propelled about the immigration of the original Bhotiya settlers from places other than the Tibetan plateau.[31] It was argued in great detail that customary arrangements would not contradict Hindu religious beliefs and practices but rather represent local and uncorrupted varieties of them. While *rambang* was uniformly discussed as a thing of the past, emphasis was placed on the egalitarian outlook that tribal institutions would in general maintain. A clan-based social organization and, still more, a good societal standing for women were considered as clear illustrations of this. Published during a time when the anti-dowry struggle gained a foothold in India, such analyses also allowed for a highlighting of such rather doubtful practices of mainstream society that never found their way into local arrangements.[32] In general, these scholarly inputs established acceptable forms of cultural objectification that in turn bolstered the overall concerns of young and educated community members to cultivate a sense of pride in what would these days be called their own 'cultural heritage'.

Outside of Kumaon, Bhotiya undergraduates and young professionals in Allahabad, an administrative service and education centre in the state of Uttar Pradesh, established the Border Student Association (BSA) and the Young Rang Club (YRC). It is noticeable that the members of these urban formations, which can be compared to university fraternities, increasingly replaced the ethnonym 'Shauka' with 'Rang'. The latter term is the self-designation used in the different languages spoken by the eastern Bhotiyas. Its public usage indicates that people started to

[31] These authors developed new hypotheses with regard to the settlement histories of their valleys, tracing them to migration flows from the ancient Indus valley civilization (Raypa 1974) or the fringes of the Roman Empire (Garbyal 1987).

[32] The practice of dowry weddings, also called *kanyadan* or 'the gift of the virgin', has undergone significant changes in India. According to the ancient classification of marriage according to the Dharmasastras, brideprice is considered as a 'demonic' (*asura*) form. Also in contemporary times people all over South Asia continue to regard it as inferior. Tambiah (1973: 69–71; cf. Sax 1991: 104–115), for instance, argues that there exists an ongoing moral pressure to transform bridewealth and brideprice into dowry. This often has a rather negative impact on women's status and survival (Srinivasan 2005: 595). Until present times the villagers in Darma express their dislike of dowry and regularly depict it as a custom that some local people adopt in order to appear ever more 'modern'.

identify themselves with pride as one distinctive branch of 'the Bhotiyas'. The members of these associations established good contacts with journalists, social activists and regional politicians to articulate their concerns. However, the central administration of Uttar Pradesh (U.P.) showed little interest in the mountain periphery and its residents, regardless of whether they were tribal or not.

To this day the Government of U.P. implemented 'Kumaon and Uttarakhand Zamindari Abolition and Land Reforms Act 1960' has left a bitter feeling amongst the Bhotiyas. The act was drafted with the intention of transferring the ownership of agricultural land to the actual cultivator (GoUP 1967). It should thus strengthen the interests of 'weak' groups, such as Scheduled Castes and Scheduled Tribes. However, until then the political leaders of U.P. had not made any recommendation for the public notification of a Scheduled Tribe, so that many Bhotiya families lost access to pastures and cultivatable land to their tenants, especially in the foothills and the Lesser Himalaya (Gerwin and Bergmann 2012: 99–100). Already under British colonial rule Bhotiyas had acquired respectable landholdings around their winter settlements and along their migratory route toward the Terai (Brown 1984: 108–113). Since trans-Himalayan trade and seasonal migration made it difficult for them to make use of their landed property during a large part of the year, they installed non-migratory (and often high-) caste Hindus as tenants in these places.

People note with bitter irony that they became officially recognized as a Scheduled Tribe only 1 year after the act was finally promulgated in 1966. Despite the fact that many appeals and requests were raised, the state government failed in granting them any compensation for the expropriated lands. Some people draw a lucid comparison: like their heroine, Old Lady Jasuli, who had to make headway as a desperate widow, they also were obliged to make a living as a tribal group bereft of its profits from cross-border commerce and landed property. The point is that Jasuli could reckon on the state, whereas those who lost their land could not. However, while such perceptions carry the plot to a strategically favorable ending, especially since they justify the need for enduring protection of the ST community, things were a bit more complex on the ground. With regard to the land reforms some households were – at least partially – able to bypass these ongoing procedures through the (temporary) cultivation of fields by themselves or with the assistance of relatives who replaced the installed tenant farmers. Moreover, wide parts of the community successfully drew from the various incentives offered by the state after 1967. This was facilitated by the various institutional frameworks, such as welfare and student organizations, which helped people to bridge the gap when shifting from a high mountain to an urban environment in pursuing work opportunities outside the agricultural sector.

In summary it can be said that Jasuli conforms to the ethic of service and sacrifice for the community that Gandhi, Vivekananda and many others have tried with various degrees of success to introduce to Indian society. Similarly to the report drafted by the KBPF, Jasuli's life story forms part of the Bhotiyas' strategic effort to participate in the "reiterative and citational practices" (Kaiser and Nikiforova 2008: 552) that had created the scale effect of *Bhot* in India. By telling the outside world about the deeds of their heroine, a young generation of politically organized

Bhotiyas asserted who they are or want to be by reproducing a certain vision of independent India as a scale effect in *Bhot*. As a result, their locality did not appear anymore as a breeding ground for 'evil' practices and 'anti-social' customs but rather crystallized as an epitome of Gandhian nationalism. The Bhotiyas, in other words, appeared as different but yet not removed from or even ignorant toward mainstream society. The problem was that various observers had disregarded or misinterpreted the Bhotiyas situation. Worse still, whilst the central government had eventually notified the Bhotiyas as a Scheduled Tribe in 1967, state officials appeared yet to be ignorant on issues relating to the tribal affairs in *Bhot*, as the controversies over the Zaminadari Abolition Act suggest.

During the 1990s, resistance against the U.P. government found ever more widespread approval from all sections of society. At that time Bhotiyas already held key positions in the local and regional administration, and they were able not only to retain, but also to expand their property in the bazaars of Dharchula and Munsiari. It was against this backdrop that the 'competing inequalities' (Corbridge 2000) between target and non-target populations of tribal reservations came to the fore. Caste Hindus suddenly staked the claim that the Bhotiyas' Scheduled Tribe designation would not be an ethnographic term that refers only to a specific group of people, but rather a scalar expression that pertains to a region including all its 'native' inhabitants.

3.6 Negotiating 'Bhotiya' as a Scalar Expression

After a devastating earthquake had hit the town of Uttarkashi in 1991, residents in the Garhwal and Kumaon Himalaya became most dissatisfied with the Government of Uttar Pradesh due to its bad handling of the disaster. Moreover, when people started to protest against the disorganized and slow relief operations, the police immediadetly used tear gas to keep them quiet. This shocking experience fuelled the movement for Uttarakhand's autonomy, a movement that met with even greater popularity in 1994. The then chief minister of Uttar Pradesh, M.S. Yadav, drastically changed the state's reservation system for the marginalized sections of society, particularly for communities designated as Other Backward Classes (OBCs).[33] Following the Mandal Commission's recommendations, Mr. Yadav greatly increased the proportion of places reserved for OBCs in government posts and educational institutions.[34] This was a well-calculated political move to win voters in the plains

[33] Other Backward Classes are generally perceived as being neither 'ritually' (like Scheduled Castes) nor 'civilizationally' (like Scheduled Tribes) deprived but as suffering from their economic backwardness. Because individual membership to a class grouping among Hindu-OBCs is predominantly defined in terms of caste (Kapila 2008: 124), the category closely interweaves issues of caste and class (Jaffrelot 2003; Chandra 2004).

[34] The Mandal Commission was a committee of five lawyers appointed in 1978 by the Government of India. Its task was to establish sound indicators for determining OBCs and to provide policy recommendations for affirmative action programs that could improve their situation. While the

of Uttar Pradesh, where the section of OBCs is rather large. In the hills, however, they only represented around 2–3 % of the total population (Mawdsley 1996: 209). It thus stands to reason that many mountain dwellers felt once again unjustly treated and passed over by the government, since this offer was of use only to a tiny proportion of the state's residents.

Ex-servicemen, students and women-groups started various protest marches in which they tied on earlier demands for autonomy and federal independence.[35] Demonstrations were violently quashed by the police (Handa 2002: 217–218). Moreover, the chief minister publicly discredited the overall movement as an anti-reservationist project. In that manner he thought "to cement his own fraying social coalition among lower castes and highlight his own social justice credentials by portraying the protestors in the hills as chauvinistic upper castes whose main concern was their opposition to affirmative action for lower castes *tout court*" (Tillin 2011: 83). This strategy, however, distorted the fact that a majority of people who participated in the protests was not against reservations but rather against any kind of policy that – in their view – "showed no real concern for, or understanding of, the needs of hill people" (Sax 2011: 172). Although the ethnic-factor did not play a major role in the movement for independent statehood, these overriding political maneuverings created new turbulences for the hills' tribal residents, especially for the Bhotiyas. Suddenly caste Hindus felt encouraged to put into question the terminological clarity of the Bhotiyas' Schduled Tribe designation, and tried to re-negotiate its 'proper' referent for their own benefits.

Only 1 year after the violent confrontations in 1994, Bhotiyas in Pithoragarh District founded the Border Bhotiya Uplift Committee (BBUC) (Dhakriyal 2004: 332). At almost the same time caste Hindus around Dharchula and Munsiari also started organizing themselves politically. Fuelled by the public controversy over the reservations system and the general neglect of the region in terms of social development, educational prospects and economic growth, these non-tribal residents felt disadvantaged in comparison to their tribal neighbors.[36] As a then still loose coalition,

commission's report was already published in 1980, it was only a decade later when the then Prime Minister of India, Vishwanath Pratap Singh, started to implement some of the suggested recommendations for reservations in the educational sector. Until present day this implementation process has not yet been completed, especially because it provoked substantial and partly violent opposition.

[35] In 1989 various left-leaning groups joined forces and founded the Uttarakhand United Struggle Association (USSS), a prime mover in the early movement for independence. After 1995, the Bharatiya Janata Party (BJP) became also involved and gradually enjoyed a large clientele. This is also evident from the fact that Uttarakhand's first Chief Minister, Nityanand Swami, was a member of the BJP. The BJP distinguished its activities from those of the various other groups involved in the independence movement by proposing Uttaranchal (meaning 'Northern Region') as the state's name.

[36] It should be mentioned that the caste Hindus backed their overall claims with reference to articles 15(4) and 16(4) of the Indian Constitution. The former article is a specification of the constitutional right to equality and says that the prohibition of discrimination shall not "prevent the State from making any special provision for the advancement of any socially and educationally backward classes of citizens or for the Scheduled Castes and the Scheduled Tribes" (CoI 2011). The

they submitted a first memorandum to India's Commission for Backward Classes (Dhakriyal 2004: 332–335). In this memorandum they demanded that the central government should consider the possibility of expanding the number of groups subsumed under the category 'Bhotiya' in order to make the Scheduled Tribe benefits accessible to a wider range of needy people in the area. While the request was dismissed, it still caused public furor, especially amongst the Bhotiyas who anticipated a growing pressure for limited state subsidies.[37]

The situation became aggravated in 1999, when non-tribal residents around Dharchula submitted a more elaborate petition to the High Court at Allahabad, Uttar Pradesh, under the name Bhotiya Scheduled Tribe Struggle Committee (BSTSC) (UHC 2004). The petitioners argued for a scalar understanding of the name 'Bhotiya', since imperial gazetteers and other public documents would consistently depict all people in the region called *Bhot* as 'Bhotiya'. This was meant to distinguish them from both the Kumaonis in the South and the Huniyas (Tibetans) in the north. Since the settlements of Dharchula and Munsiari have always been considered as a part of that particular region, all native residents should by implication also be included under the public notification that specified 'Bhotiya' as Scheduled Tribe. In their view the government had already followed such an approach in the case of the Jaunsari of Garhwal, since the 'natives' of Jaunsar would have with good reason and irrespective of any further criteria, such as caste, been recognized as Scheduled Tribe. Why, so it was then asked, was such an approach not adopted in the case of *Bhot*?

Non-tribal leaders thought that state actors had most probably become dazzled by the dominant presence of 'the Bhotiyas'. For a long time the interests of the latter would have overshadowed the concerns and poor living conditions of other resident communities in the area. One can easily imagine that such accusations were a hard blow for the Bhotiyas, even more so when caste Hindus in Dharchula referred to themselves as Bhotiya Brahmin or Bhotiya Thakur (Rajput). The BBUC tried to counteract these actions through the consolidation of intra-community linkages and the establishment of alliances to other tribal groups in the hills. Only 1 month before Uttarakhand (named Uttaranchal between 2000 and 2002) received its federal independence in November 2000, this endeavor achieved a major success with the founding of a statewide representative organ, the Uttaranchal State Tribal Organization (USTG) (Dhakriyal 2004: 332–333).

This organization, in which Bhotiyas played a leadership role, was clearly geared toward the maintenance and improvement of state programs for the protection and promotion of STs. It advocated the reservation of seats in parliament and legislative assemblies; their inclusion in the state's delimitation commission (*parisiman*); the

---

latter article further says that also in matters of public employment the right of not being discriminated shall not prevent "the State from making any provision for the reservation of appointments or posts in favour of any backward class of citizens which, in the opinion of the State, is not adequately represented in the services under the State" (CoI 2011).

[37] These subsidies are allocated, for instance, through the 'Border Area Development Program' (BADP) and the 'Tribal Sub-Plan' (TSP) (Gerwin and Bergmann 2012: 101).

availability and transparent allocation of reserved jobs; the installation of a ministerial office for ethnic affairs; the determination of quota allotments in local governmental bodies (*panchayats*); the right to a Tribes Advisory Council (TAC)[38]; and the conversion of tribal-inhabited areas, such as Munsiari and Dharchula, into separate districts or Scheduled Areas. While most of these issues are still under consideration, the Dharchula constituency – comprising both Dharchula and Munsiyari Tehsil – was provisionally declared as a ST-seat for the first assembly elections in 2002. During the election campaign, the struggles over the Bhotiya's Scheduled Tribe designation took a new direction.

A number of Bhotiya leaders announced their candidacy with great confidence, but many caste Hindus were unwilling to support any of them. In a highly politicized atmosphere they pushed an independent candidate from another tribal community settled within the constituency, but hitherto completely absent from its political scene. It came as a great surprise to many observers when Mr. G.S. Rajwar, a member of the Raji tribe, who had until then made his living as an unskilled construction worker, edged out all other competitors with a sizeable lead, viz. 42.52 %.[39] What else could the disappointed Bhotiya candidates do, other than grasp the nettle and congratulate their tribal brother? Publicly the outcome was sold as an indicator of the success of the state's tribal policy. Finally, it was said, one of the most 'primitive' peoples of the state had been lifted out of isolation and placed in a position to participate in development and progress. Beneath the smooth surface, however, there was seething unrest. Shortly after Mr. Rajwar was officially appointed, the BBUC published a protest letter via the Uttarakhand unit of the Backward and Minority Community Employees' Federation (BAMCEF) that was addressed to India's National Commission for Scheduled Tribes (BAMCEF reprinted in Dhakriyal 2004: 336–338).

In this letter the Bhotiya authors depicted the whole strategy of their non-tribal competitors as an elite conspiracy: to recognize members of the regional caste society as 'Bhotiya' would not only be unconstitutional but, in the long run, also counterproductive. The 'real' tribe (*janajati*), so the worry, was likely to disappear altogether from the map of India once it was possible for the members of a general caste (*jati*) to appropriate the benefits of reservations that were meant for the weak sections of society. If the government approved such a development, so the authors wrote, it would turn the initial spirit of tribal reservations on its head. State bureaucrats were therefore called upon to protect the interests and ensure the rights of the

[38] The Tribes Advisory Councils shall advise the state governor in matters relating to the welfare and advancement of the state's tribal population. They are principally envisaged for federal states that have a Scheduled Area in their territory. If a state does not fulfill this provision, the president of India can still direct federal states to establish such a TAC, as it happened in Uttarakhand (GoIE 2010: 263). The Presidential Order proclaims that a 19-member TAC is to be constituted in Uttarakhand. However, the implementation of that order was delayed due to political power plays in the highest echelons of the Uttarakhand government (Garhwal Post 2012).

[39] The Raji have traditionally lived as hunter-gatherers in the forests of the Lesser Himalaya of eastern Kumaon and adjacent regions in Nepal (Fortier 2002). They are also referred to as Van Raja or Van Rawat, which translates as 'kings of the forest'.

tribal population, or expect a massive protest movement. Information from colonial sources served to underline this position. A passage from E.T. Atkinson's 'The Himalayan Gazetter', which is widely sold in the bookshops of Uttarakhand, turned out as a most relevant point of reference. Atkinson (1973 [1882]: 84, italics mine), following an earlier remark by J. Strachey, had argued that

> the limits of Bhot in these districts [viz. Garhwal and Kumaon] cannot be strictly defined, *for the term is an ethnographical rather than a geographical expression*, and signifies rather the tract inhabited by the Bhotiyas which, too, has varied for time to time, than a country of which any positive boundaries can be named.[40]

The BBUC acknowledged that the popular name of the region inhabited by the Bhotiyas, viz. *Bhot*, had once influenced the British to officially adopt the ethnic designation 'Bhotiya'. However, it would be rather odd, so they continued, to assume that everyone living in *Bhot* is automatically a Bhotiya since even the British had eventually recognized that the referent of the term could only be defined ethnographically. Moreover, it would be evident that the "culture, language, deities, festivals, customs, rites, foods and drinks [of the caste Hindus] are absolutely different from those of the Bhotiyas" (BAMCEF 2002, reprinted in Dhakriyal 2004: 337). Some months later, namely in May 2003, the Uttaranchal High Court also deposed the still pending petition issued by the BSTSC (UHC 2004). The judgment was based on a legal precedent established by the Supreme Court (SCI 2000: 4, italics mine) 2 years earlier, which stipulated that

> [w]hether a particular caste or a tribe is Scheduled Caste or Scheduled Tribe as the case may be within the meaning of the entries contained in the Presidential Orders issued under clause (1) of Articles 341 and 342, *is to be determined looking to them as they are*. Clause (2) of the said Articles does not permit any one to seek modification of the said orders by leading evidence that the caste / tribe (A) alone is mentioned in the Order but caste / tribe (B) is also a part of caste / tribe (A) and as such caste / tribe (B) should be deemed to be a scheduled Caste / Scheduled Tribe as the case may be. *It is only the Parliament that is competent to amend the Orders* issued under Articles 341 and 342.

The verdict provides for a *modus operandi* on how to decide whether a particular group is part of an officially recognized tribe, such as the Bhotiyas. This is to be done by verifying the inclusion of the concerned group's name in the original presidential order. If that is not the case, modifications can only be made through an act of the Indian Parliament. Since the presidential order from 1967, which recognizes 'Bhotiya' as a Scheduled Tribe, did not subsume any other group-name under that category, the BSTSC lost the case. However, its leading members interpreted the High Court's decision differently, namely as proving their case right and containing a concrete directive for administrative action. According to their interpretation the judgment declared that the meaning of the names contained in the presidential order "is to be determined looking to them as they are" (SCI 2000: 4). State administrators would therefore be judicially obligated to acknowledge that the term 'Bhotiya' *is* a scalar expression. As such it would pertain to a region, including all of its 'native'

[40]The article to which Atkinson refers is Strachey (1853).

residents, and not to only one specific group that claims to be culturally distinctive and conterminous with *Bhot*.

About 2 months after the judgment was made, members of the BSTSC assembled and requested the district magistrate of Pithoragarh to start implementing the court's order along these lines. But the district magistrate only forwarded the request to the then chief secretary of Uttarakhand, Dr. R.S. Tolia, who did not see the necessity in taking any action as well. The BSTSC immediately suspected that a targeted manipulation was taking place and that Dr. Tolia, himself a Bhotiya, might be biased. Without further ado they filed another petition demanding the punishment of both the district magistrate of Pithoragarh and the chief secretary of Uttarakhand for their willful disobedience of the High Court's order. However, in August 2004 the Uttarakhand High Court also deposed this second petition and agreed with the accused persons that the original judgment did indeed contain no directives for action. The aforementioned decision by the Supreme Court was again quoted to further underline the fact that neither the district magistrate nor the chief secretary would in any case have been entitled to intervene in matters relating to the presidential orders. The Indian Parliament alone has the authority to answer questions about who is, or is not, included under the name of a Scheduled Tribe. An additional reference was given to Dr. B.R. Ambedkar (CAD 1949: 1637) who had pointed out to the Constituent Assembly

> that once a notification has been issued by the President, which, undoubtedly, he will be issuing in consultation with and on the advice of the Government of each State, thereafter, if any elimination was to be made from the List so notified or any addition was to be made that must be made by Parliament and not by the President. The object is to eliminate any kind of political factors having a play in the matter of the disturbance in the Schedule so published by the President.

In other words, only the parliament is allowed to modify the presidential orders in order to prevent state politicians from using such incidences as a means to generate voter confidence. With that judgment the case was for the time being shelved in favor of the Bhotiyas, at least in court.[41] Encouraged by the verdict especially the educated, affluent and urban dwelling community members made further efforts to accentuate their distinctiveness as an ethnic group. The Rang Welfare Association (RWA) assumed a special role in that regard. The association was founded in Lucknow in 1989 to support students and provide rural-urban migrants with a common platform. At the time of Uttarakhand's independence various branches had been established in other cities, such as Haldwani and Dehra Dun. From 2004 onwards the RWA institutionalized an annual general meeting for the whole community. These meetings serve to discuss current affairs with invited state officials as well as to stage performances that in the association's own words are meant

[41] The High Court also emphasized that whilst the petitioners pointed out that the Legislative Assembly (*Vidhan Sabha*) of Uttarakhand had given its recommendation to notify the region in question as a Scheduled Area, they had wrongly accused Dr. Tolia of preventing its implementation. In actual fact, however, Dr. Tolia had himself signed a letter of pursuance thereof that he sent to the Ministry of Tribal Affairs in 2001 (UHC 2004: 3).

"to preserve the individuality of our culture, introduce the new generation to our colorful society, re-install the rung [sic] values, and for most to bring together rungs from various regions of the globe" (Rungmung 2013).

In April 2008 I documented one such meeting in Dharchula, which allowed me to witness how people now try to publicly foreground an ethnographic understanding of the name Bhotiya (Bergmann et al. 2011). Prakash Pant, then Uttarakhand's minister for Tourism, Culture and Pilgrimage, attended it as honorary guest. The ceremonial opening was followed by intensive discussions about the opportunities and challenges for increasing the region's tourist influx. In that regard Mr. Kutiyal, the aforementioned owner of the local trekking agency, was also awarded for his pioneering work. But only toward the conclusion of that meeting, which was on the evening of the second day, every single seat was taken. Hundreds of attendees, tribals and non-tribals alike, eagerly awaited the cultural competition between dancers and singers of the various Bhotiya valleys. A group coming from the Mana and Niti valleys in Garhwal received the most intense public acclaim. Their performance centered on the former trans-Himalayan trade, particularly on the relationship with Tibetan business partners.

Under British rule the Bhotiyas had popularized the half-truth that the Tibetans regard commensal relations as a necessary requirement for commerce, and that Indian traders had therefore to disregard Hindu rules about touch pollution and beef eating (Srivastava 1966: 181). Besides reasons of logistical feasibility and language competence, this kept potential high-caste Pahari competitors even more effectively away, thus strengthening the Bhotiyas' monopoly. But when influential community members sought public acceptance as high-caste Hindus during late colonial rule, the impact of this preconception proved rather disadvantageous and eventually stimulated a change in strategy towards tribal status (Srivastava 1966). However, the ethnonym 'Bhotiya' still carries the connotations of beef-eating and Tibetan-like cultural practices, and is therefore "not liked by the various communities to whom it is applied" (Tolia 2010: 50). The dance group re-redefined the apparent tension transmitted by the ethnonym in various scenes that were elucidated by the following sequences in Hindi:

> The glaciers that confine this picturesque valley, this beautiful land – wow! How lovely it is! How lovely it is! The holy abode of saints! And the name of this heaven-like land is Niti-Mana, a land as colorful as *Shiva's* Kailash.
>
> The long, monotonous and risky journey to that far away and remote land does finally come to an end. It is drawn to a close in a long procession. Finally we have arrived back in our village! 'May everyone be happy, may everyone be free from illness!' Come home in cheerful spirits! Let us remember, worship and pray our *ishta devi-devata* [favored deities].
>
> The worship of our *ishta devi-devata* has been concluded. Hey brother! Hey uncle! What! Are you still sitting around enjoying your *prasad* [food remnants from a deity]? Don't we have to complete our chores? Very soon we must decamp and continue our way toward Tibet for doing business [the bleating of sheep and goats rises in the background]!
>
> In this way our traders load all their goods and belongings on horses and mules, on sheep and goats before they set off for Tibet [again the sounds of animals can be heard, which are gradually drowned out by strong winds]. They need to cross the stormy passes, face danger-

> ous obstacles and cope with demanding hindrances. Our traders know it very well. 'We will go to Tibet! We will go to Tibet!' This is how our journey continues.
>
> Finally the long waiting has come to an end. The day of our re-union across the border of two countries is looming. Two cultures will come together and two friends will share their feelings. But what kind of meeting is this? Though it is a commercial activity, the whole affair is less about the exchange of commodities than about the exchange of feelings. The Tibetan trade is all about this! The Tibetan trade is all about this!
>
> 'Listen friends, next year we will meet you again!' With this promise our traders come back. They face the strong winds and the dust on the way to the Niti-Mana valley, on the way to the Niti-Mana valley. When the traders return to their village safely, they spread happiness and joy among the people. Men and women, young and elderly, everyone becomes excited. They are all wondering what the traders might have brought for them: is it a bracelet, an anklet, or some necklace?
>
> The constant blowing of the wind, the twittering of birds, the sound of the rivers and the roaring of fountain-like waterfalls, it seems that all creation has come to a halt. We have noticed that time is not in favour of our fortune, of our prosperity and our trading activities. Commerce stopped and our economy is tumbling. What will happen to the next generation? What will happen now? What will happen now? Think about it, friends! What will happen now? What will happen now?
>
> Though a border limits our cultures, we are still willing to exchange feelings. Friends, the desire to meet you is still alive in our hearts! Yet we know that the Tibetan trade has been engraved on the pages of history like a bygone event.
>
> But there is hope that on these pages of history our trade and our emotional exchange will merely pause for breath. These former events might rise, when the course of history changes side again. Please forgive us, friends, that for the moment we are unable to keep our promise of meeting you. We are unable to keep this promise; we are unable to keep it; we are unable to keep it!

There is, of course, nothing new about the idea that the Bhotiyas' ethnicity is closely linked to their involvement in trans-Himalayan trade (Brown 1987). Nevertheless, the performance indicates an interesting turn. In earlier times the ritually established trade partnerships served the Bhotiyas to strengthen their lucrative commercial monopoly vis-à-vis caste Hindus. While these partnerships had lost their economic importance even before the Sino-Indian border war in 1962, they are nowadays again doomed significant – especially on stage – for another reason: they foreground an ethnographic understanding of the name 'Bhotiya', allowing the people whom the state has recognized as such to connect themselves with a common past that is not at all considered disreputable, as it was the case with regard to *rambang*, but rather celebrated and experienced as a source of pride. This past helps to strengthen the conviction amongst the Bhotiyas that the benefits they receive as a Scheduled Tribe accrue to them rightfully and that caste Hindus are excluded from these benefits for good reasons. Caste Hindus would have always looked down upon the Bhotiyas because they maintained emotional bonds with the Tibetans. This, then, reveals their current claims for public recognition as Bhotiya as nothing else than an act of hypocrisy and deceit.

While this process of heritagization has enhanced the Bhotiya's stance in the struggle over their Scheduled Tribe designation, the community still faced a serious political setback in the course of Uttarakhand's third assembly elections in March 2012.[42] For the purpose of these elections the boundaries of the legislative constituencies in Uttarakhand were redrawn and, as a direct result, Dharchula lost its ST-reserved seat. Tribal residents from all over the state were rather disappointed with this decision of the state's delimitation commission (*parisiman*).[43] Mr. Rajwar was defeated after 10 years in power by a non-tribal candidate from the Indian National Congress (INC). This candidate adopted a new strategy by taking a strong stand in declaring his constituency as a 'backward area' and for getting 183 castes conferred to the OBC-status (Tribune 2012).[44] However, the dispute over the question of what defines and who actually is a 'Bhotiya' has remained on the table of government officials at the high echelons of the Indian state.

In August 2012, the lower house of the Indian Parliament (*Lok Sabha*) discussed the issue of how to proceed with the large number of proposals dealing with the inclusion of specific castes or tribes into the list of Scheduled Tribes, submitted from individual states to the Union Government over the last 3 years (GoI 2012). The list contains 243 proposals out of which 2 are from Uttarakhand, these are: the demand to add the categories 'Bhotiya Brahmin' and 'Bhotiya Rajput' to the public notification from 1967, as well as the demand to classify all native residents of Dharchula and Munsiari Tehsil as Scheduled Tribe. In line with the former verdicts of the Uttarakhand High Court the *Lok Sabha*, however, stressed that any amendment could only be accomplished through an act of the Indian Parliament. Moreover, it pointed out that a timeframe couldn't be assigned to decisions of individual proposals. In other words, quick results in that direction seem rather unlikely and without any real chance of succeeding. In the meantime, however, the Bhotiyas followed on from their success of the previous decades. Only a few days after the *Lok Sabha* debate had taken place in New Delhi, an exceptionally high number of people crowded the tiny mountain settlement of Danthu in the Darma valley.

Bhotiyas from all the different valleys had come, many from far-away places in India or even from abroad, to participate in the Danthu *mela*. This annual fair was launched by the BYWA during the late 1980s.[45] In it all the villages compete for a

---

[42] The process of heritagization is enhanced through various forms of public display, ranging from print magazines to community websites, and from social media platforms to museum installations. By now there are dozens of platforms in the Internet maintained by different sections of 'the Bhotiyas', from websites and blogs to representations on Facebook and Twitter. The Rang Welfare Association also publishes an annual magazine called *Amtikar*, which literally translates as 'road map', to publicly disclose and monitor communal affairs, interests and developments. External visitors to the region can further frequent both the Tribal Heritage Museum in Munsiari and the Rang Museum in Dharchula to get an insight into the history and culture of 'the Bhotiyas'.

[43] Whilst the number of ST-reserved constituencies had been reduced to 2 (Chakrata and Khatima), the number of SC-reserved constituencies was simultaneously raised from 12 to 13.

[44] One needs to keep in mind that the movement of the caste Hindus still consists of several sub-factions with slightly different agendas, which I have left out in my discussion.

[45] In the Byans valley a similar *mela* was already introduced in the 1970s in the village Manila. There as well the idea was to draw out a positive image of traditional Bhotiya culture, one that would not anymore mislead people to forecast a 'gloomy prospect' for those engaging in it.

shield bearing Old Lady Jasuli's name by staging traditional lore, music and dance. Normally it is only the migrating Darmani who participate in this 1-week lasting event. That so many people came in 2012 was because the community's annual general meeting followed the *mela*. It was for the first time that this meeting took place in a high altitude settlement. By having chosen the Darma valley as a venue, the organizers thought it would gain in symbolic power. This became clear when the federal state's information commissioner, Mr. N.S. Napalchyal from the Byans valley, ceremonially inaugurated the meeting at a place that is nowadays deemed to be iconic for the Bhotiyas' tribal identity, namely in front of Old Lady Jasuli's statue. Her character is deemed to be conclusive evidence for the fact that the name 'Bhotiya' is first and foremost an ethnographic and not a scalar expression. Bhotiya-ness, in other words, has not so much to do with people's place of residence as with their ethic of service, sacrifice for the community and cultural distinctiveness.

3.7 Coda

A question that remains open is why the Bhotiyas have, despite all controversy, never made any serious move to change the ethnonym that is used for them in the official jargon. Outside observers have long felt that the category 'Bhotiya' is a misnomer (Brown 1992), and those subsumed under it prefer other names, such as Shauka or Rang, in their daily conversation. Based on the material presented in this chapter, I contend that people have had good reason to maintain the status quo. First, there appears to be no collectively agreed alternative. Second, while the designation 'Bhotiya' is felt to bear some negative connotations, these are counterbalanced by various and officially recognized positive attributes. Third, since the term was instrumentally attractive to powerful interests amongst state actors, especially but not exclusively under British rule, it helped to confer political weight and legitimacy to the claims and aspirations of the Bhotiyas.

On the eve of India's independence the members of the newly founded KBPF assembled representations that colonial administrators had coined about the Bhotiyas, especially in terms of gender relations. The Federation not only claimed that the 'Culture of the Bhotiyas' was distinct from other regional groups, but also maintained that its people would face serious difficulties in adjusting themselves to the emergent political setup of postcolonial India. Similar to British officials, the Federation referred to gender issues for distinguishing *Bhot*, or the region inhabited by the Bhotiyas, from surrounding areas (with its mainstream Hindu or Buddhist residents) as well as for highlighting the changes that had already occurred within *Bhot*. A customary institution, viz. *rambang*, was taken as a kind of litmus test for estimating the varied degrees of success according to which the different valley communities would have already realized their 'duty' toward the nation. *Rambang* allowed unmarried men and women to interact rather freely, a circumstance that the Federation considered as highly problematic. However, the alleged primitiveness of those community members who openly practiced *rambang*, most notably the

Darmani, not simply compromised the honor, reputation and esteem of this political elite but also legitimized the demand for special statutory safeguards and constitutional protection as a tribal minority.

The Federation argued that the Bhotiyas may have travelled great distances as trans-Himalayan traders, but they would yet have always remained far from India's mainstream society. Former unsuccessful attempts to achieve acceptance within the regional caste system were thereby re-interpreted as a sign that only underlined their status as what Ghurye (1943) called 'primitive Hindus'. Outmoded customs and a lack of education would have prevented them from advancing endogenous development and assuming social responsibility. In other words, the Federation tailored the socio-spatial imaginary of a tribal people who were afflicted by the 'evils of ignorance' in the neglected margins of the state. This imaginary enabled them to both accentuate the Bhotiyas' tribalness and to produce *Bhot* (or the region inhabited by the Bhotiyas) as a scale effect in independent India. With the awarding of Scheduled Tribe status in 1967, this effect of scale became, at least temporarily, "sedimented and 'fixed'" (MacKinnon 2011: 31), particularly since the Bhotiyas could then benefit from India's tribal policy.

However, it further became clear in this chapter that the strategy pursued by the KBPF also reinforced rather unequal power relations within the ST community. The public defamation of *rambang*, for instance, allowed an aspiring circle of a modernist and reformist Bhotiya elite to substantiate the powerful position that this creamy layer had already seized under British colonial rule. In requesting government officials to impose a legal ban on allegedly 'anti-social' customary institutions such as *rambang*, the Federation espoused a limitation of their women's freedom whilst enlarging its own influence within webs of relations at 'higher' levels of activity, especially the national. Only after the Sino-Indian border war in 1962 and after the Bhotiyas had officially been notified a Scheduled Tribe in 1967, those people whom the Federation had confined to the 'lower' scale of an isolated mountain region, particularly the Darmani, found ways and means to steer these negotiations in an entirely new direction.

A young generation of Bhotiyas organized themselves within various institutional frameworks, such as welfare and student associations. The popularization of Old Lady Jasuli's life story evolved as a crucial building block of their endeavors. Through her character they could convey an empowering sense of identity by reproducing independent India as a scale effect in *Bhot*. As a de-sexed widow who is committed to the common good Jasuli conforms to the ethic of service and sacrifice for the community that Gandhi, Vivekananda and many others have tried with various degrees of success to introduce to Indian society. In other words, by telling the outside world about the deeds of their local heroine, these community members challenged and transformed the political power dimensions of inherited socio-spatial imageries. Jasuli's life suggests that it was not them who failed to recognize their 'duty' within a wider societal context, but rather that various observers (including the KBPF) would have ignored the forms of social commitment that had emerged in *Bhot* already under late British rule.

The actors involved in this movement encouraged local villagers to take pride in their tribalness, for instance by organizing various cultural events, and to make use of tribal reservations as a way to access formal education. The increasing number of Bhotiyas who then managed to pursue career opportunities outside the agrarian sector did not perceive their culture as the root of all evil that was in need of 'mental, moral and cultural uplift' but rather made efforts to keep alive the local fields of practice to which they attributed their own success. But whereas the size of a well-educated and urban-dwelling Bhotiya middle-class increased steadily, the caste Hindus of the area felt evermore disadvantaged. This was especially so because the Government of Uttar Pradesh, of which Uttarakhand then formed a part of, did as little as possible in developing this mountain region. When the Uttarakhand movement for independent federal statehood gradually gained momentum in Garhwal and Kumaon during the mid-1990s, it was the question of tribal reservations that dominated the day-to-day political dealings in the region.

Non-tribal residents around Dharchula and Munsiari founded their own political associations through which they exerted increasing pressure on the Bhotiyas. Their demands touched a rather sensitive issue, namely the sense and referent of the designation 'Bhotiya'. Ultimately they propelled the argument that the Bhotiyas' Scheduled Tribe designation is a scalar expression: all the 'natives' of the *region* that British officials had labeled as *Bhot* should be included under the ST notification from 1967. In facing this movement the Bhotiyas established new alliances, both amongst themselves and with other tribal communities in Uttarakhand. They sensed that any questioning from their side of the appropriateness of the term 'Bhotiya' equaled to, as Tolia (2010: 50) puts it, "the opening of a Pandora's box". People suspect that any move in that direction might have pulled the rug of government protection right out from under their feet. A new or expanded nomenclature was likely to encourage new forms of identifying self and other. People seemed aware of the fact that new names not only make up new kinds of people but also that they can easily acquire a dynamic, and perhaps even more difficult to control life of their own (Kapila 2008: 130; Hacking 1995). Be that as it may, the Bhotiyas were well advised to amplify their efforts under the pretext of the already established classification, because it allowed them to effectively counter the caste Hindus' demand.

The material presented in this chapter thus stresses the importance of considering the implications of a contentious scalar politics in processes of ethnic identification, particularly in connection with such macro-institutional transformations as the shift from British imperial rule to India's nation-state sovereignty or the creation of new states within that country's federal system of government. The Bhotiyas have leveraged the performativity of scale through various social practices and narratives to effectively voice their demand for tribal status in postcolonial India. There are two main lessons to be learnt from this analysis. First, when considering scales as performatives they need to be understood not as being made and re-made but rather as being practiced within the strategies that actors deploy whilst pursuing their various agendas. However, these strategies and the very real effects they produce do not occur in a vacuum but rather in a "multiscalar discursive field" (Kaiser and

Nikiforova 2008: 544). The recent attempts of caste Hindus demonstrate, for instance, that legal regulations can hinder (or in other cases promote) the efficacy of any such performance. Second, although notions of hierarchy do play a role in a contentious scalar politics, social relations of empowerment and disempowerment do not evolve along a clear-cut hierarchical order. Rather, as the Bhotiyas' struggle for Scheduled Tribe status has illustrated, such relations evolve rather in the shape of what Brenner (2001) has called 'mosaics', or power geometries that are constantly shaped and re-shaped through the "interaction between inherited scalar structures and emergent regulatory projects and strategies" (MacKinnon 2011: 25).

Until the present time the Government of India maintains that tribal societies are egalitarian and more or less undifferentiated. Moreover, it is widely believed that the emergence of a tribal elite or middle class – if it exists at all – has only been made possible through the benefits and incentives that the state has on offer to its scheduled populations. Corbridge (2000: 64) has criticized both views and argued instead that the Indian reservation system "has not brought a tribal middle class into existence" but rather consolidated it, and that the introduction of reservations has neither served only a pre-existing elite nor only selfish power interests. A similar conclusion can be drawn with regard to the Bhotiyas. Not only was there a well-established Bhotiya elite before India's independence, but the community's public notification as a Scheduled Tribe also encouraged those whom that elite had confined to the 'lower' scale of an isolated tribal region to take advantage of this novel status and thrive. Nowadays the former underdogs, viz. the Darmani, don't feel overlooked or misrecognized anymore. They rather set the tone when it comes to articulate the cultural dignity and cultural heritage of the Bhotiyas. A song that has recently been composed and popularized at official community gatherings illustrates this quite nicely:

Rang Raju [the land of the Rang] is an inseparable part of India,
Its existence is glorious, immortal.
This is the fruit of our fathers' labors
And why we all say with our heads held high:
We are Rang! We are Rang! We are Rang!
Every son of a Rang, every single Rang's daughter says
We are Rang! We are Rang! We are Rang!

Our forefathers' life was forever linked
to crossing the passes into Tibet.
They traversed remote forests, icy rivers and streams,
those hundreds of miles from Tanakpur.
By crossing the passes year after year,
their existence, like that of the sun and the moon,
shines brightly and is deathless through untiring efforts.
And so we all say with our heads held high:
We are Rang! We are Rang! We are Rang!
Every son of a Rang, every single Rang's daughter says
We are Rang! We are Rang! We are Rang!

Our forefathers, in the struggle for life,
made the earth their bed and the sky their quilt.
They ate what was there, both cooked and uncooked!
They spoke all the languages, Tibetan, Kumaoni, Nepali:

they made every language their own.
They changed stone into silver and soil into gold,
Their lives were full of glory and honor,
as bright as silver and as pure as gold.
And so we all say with our heads held high:
We are Rang! We are Rang! We are Rang!
Every son of a Rang, every single Rang's daughter says
We are Rang! We are Rang! We are Rang!

While such songs are particularly favored by urban-dwelling community members, the fact remains that despite all the changes described in this chapter, there are still a considerable number of Bhotiyas who maintain a 'traditional' lifestyle. To paraphrase the song, these people do still traverse remote forests, icy rivers and streams, they do still make the earth their bed and the sky their quilt during seasonal migration, they do still speak many languages, and they do still turn soil if not in gold so at least into highly valued crops. What is it that attracts or constrains them to dwelling in altitudes that outsiders – and even most of their urban relatives – regularly consider as beautiful, but onerous as a home? What are the fields of practice to which migratory people are committed? What are the joys and sorrows they experience in everyday life? And not least, how does their ongoing engagement with each other and the environment look like? These are the questions that I deal with in the following chapter.

References

Agarwal B (1998) Widows versus daughters or widows as daughters? Property, land, and economic security in rural India. Mod Asian Stud 32(1):1–48

Aggarwal R (2004) Beyond lines of control: performance and politics on the disputed borders of Ladakh. Duke University Press, Durham

Ambedkar BR (1917) Castes in India: their mechanism, genesis and development. Indian Antiquary 41:81–95

Atkinson ET (1973 [1882]) The Himalayan gazetteer, vol III, part I. Natraj Publishers, Dehra Dun [originally published under the title 'The Himalayan districts of the north western provinces of India']

BAMCEF [Backward and Minority Community Employees' Federation] (2002) Protest letter to India's national commission for scheduled tribes, dated 08.09.2002. Reprinted in: Dhakriyal DS (2004) Himalayi Sauka sanskritik dharohar, vol I. Takshila Prakashan, New Delhi, pp 336–338.

Barnard A, Spencer J (1996) Culture. In: Barnard A, Spencer J (eds) Encyclopedia of social and cultural anthropology. Routledge, London, pp 136–143

Barth F (1969) Introduction. In: Barth F (ed) Ethnic groups and boundaries: the social organisation of culture difference. George Allen and Unwin, London, pp 9–38

Barth F (1978a) Introduction. In: Barth F (ed) Scale and social organization. Universitetsforlaget, Oslo, pp 9–12

Barth F (1978b) Scale and network in urban western society. In: Barth F (ed) Scale and social organization. Universitetsforlaget, Oslo, pp 163–183

Bates C (1995) Race, caste and tribe in central India. The early origins of Indian anthropometry. In: Robb P (ed) The concept of race in South Asia. Oxford University Press, New Delhi, pp 219–259

Bauman Z (2001) The great war of recognition. Theory Cult Soc 18(2–3):137–150
Baviskar A (2005) Adivasi encounters with Hindu nationalism in MP. Econ Pol Wkly 40(48):5105–5113
Bayly S (1999) Caste, society and politics in India from the eighteenth century to the modern age. Cambridge University Press, Cambridge
Benhabib S (2002) The claims of culture: equality and diversity in the global era. Princeton University Press, Princeton
Bergmann C, Gerwin M, Nüsser M, Sax WS (2008) Living in a high mountain border region: the case of the 'Bhotiyas' of the Indo-Chinese border region. J Mt Sci 5:209–217
Bergmann C, Gerwin M, Sax WS, Nüsser M (2011) Politics of scale in a high mountain border region: being mobile among the Bhotiyas of the Kumaon Himalaya, India. Nomadic Peoples 15(2):104–129
Bhargava R, Bagchi R, Sudarshan R (eds) (1999) Multiculturalism, liberalism and democracy in India. Oxford University Press, New Delhi
Boas F (1896) The growth of Indian mythologies: a study based upon the growth of the mythologies of the north pacific coast. J Am Folk 9(32):1–11
Boas F (1937) Part two: the interrelation of cultures. The diffusion of cultural traits. Soc Res 4(1/4):286–298
Brenner N (1998) Between fixity and motion: accumulation, territorial organisation and the historical geography of spatial scales. Environ Plan D Soc Space 16(4):459–481
Brenner N (2001) The limits to scale? Methodological reflections on scalar structuration. Prog Hum Geogr 25(4):591–614
Brown CW (1984) The goat is mine, the load is yours: morphogenesis of 'Bhotiya-Shauka', U.P., India. Lund University, Department of Social Anthropology, Lund
Brown CW (1987) Ecology, trade and former Bhotiya identity. In: Raha MK (ed) The Himalayan heritage. Gian Publishing House, New Delhi, pp 125–138
Brown CW (1992) What we call 'Bhotiyas' are in reality not Bhotiyas: perspectives of British colonial conceptions. In: Joshi MP, Fanger AC, Brown CW (eds) Himalaya: past and present, vol II. Shree Almora Book Depot, Almora, pp 147–172
Browne ECM (n.d.) Among the Bhotiyas and their neighbors. Nichols Press, Lynn
Budhwar K (2010) Where gods dwell: central Himalayan folktales and legends. Penguin, New Delhi
Business Standard (2011) Uttarakhand steps up tourist initiatives, August 16
CAD [Constituent Assembly Debates] (1949) Constituent assembly debates official report, vol 9. Lok Sabha Secretariat, New Delhi
Chandra K (2004) Why ethnic parties succeed: patronage and ethnic head counts in India. Cambridge University Press, Cambridge
Chatterjee P (1989a) The nationalist resolution of the women's question. In: Sangari K, Vaid S (eds) Recasting women: essays in colonial history. Kali for Women, New Delhi, pp 233–253
Chatterjee P (1989b) Colonialism, nationalism, and colonized women: the contest in India. Am Ethnol 16(4):622–633
Cohn BS (1987) An anthropologist among historians and other essays. Oxford University Press, New Delhi
CoI [Constitution of India] (2011) Constitution of India, updated up to ninety-seventh amendment act, 2011. Government of India, Ministry of Law and Justice. http://lawmin.nic.in/olwing/coi/coi-english/coi-indexenglish.htm. Accessed 15 Oct 2015
Corbridge S (2000) Competing inequalities: the scheduled tribes and the reservation system in India's Jharkhand. J Asian Stud 59(1):62–85
Das JC, Raha MK (1981) Divergent trends of transformation among the Kumaon Bhotias. In: von Fürer-Haimendorf C (ed) Asian highland societies in anthropological perspective. Sterling Publisher, New Delhi, pp 250–265
Dhakriyal DS (2004) Himalayi Sauka sanskritik dharohar, vol I. Takshila Prakashan, New Delhi

Dirks N (2001) Castes of mind: colonialism and the making of modern India. Princeton University Press, Princeton
Fisher WF (2001) Fluid boundaries. Forming and transforming identity in Nepal. Columbia University Press, New York
Fortier J (2002) The arts of deception: verbal performances by the Raute of Nepal. J R Anthropol Inst 8(2):233–257
Fraser N, Honneth A (2003) Redistribution or recognition? A political-philosophical exchange. Verso, London
Galanter M (1984) Competing equalities: law and backward classes in India. Oxford University Press, New Delhi
Gandhi MK (1966) The collected works of Mahatma Gandhi, vol 21 (August–December 1921). The Publications Division, Ministry of Information and Broadcasting, Government of India, New Delhi
Gandhi MK (1969) The collected works of Mahatma Gandhi, vol 34 (June–September 1927). The Publications Division, Ministry of Information and Broadcasting, Government of India, New Delhi
Gandhi MK (1971) The collected works of Mahatma Gandhi, vol 46 (April 16–June 17). The Publications Division, Ministry of Information and Broadcasting, Government of India, New Delhi
Gaonkar DP (ed) (2001) Alternative modernities. Duke University Press, Durham
Garbyal P (1987) Shauka ka sampurn itihas. King Press, Nainital
Garhwal Post (2012) BJP has anti-tribal stance, February 24
Gerwin M, Bergmann C (2012) Geopolitical relations and regional restructuring: the case of the Kumaon Himalaya, India. Erdkunde 66(2):1–17
Ghurye GS (1943) The aborigines – 'so-called' – and their future. Gokhale Institute of Politics and Economics, Poona [published in 1959 as: *The scheduled tribes*. Popular Prakashan, Bombay]
GoI [Government of India] (2001) 10th five year plan, 2002–2007, vol 2, sectoral policies and programmes. Government of India, New Delhi
GoI [Government of India] (2012) Lok sabha: inclusion of castes into SCs and STs category. Government of India, New Delhi
GoIE [Gazette of India Extraordinary] (2010) The tribes advisory council (state of Uttarakhand), order 2010. Government of India, New Delhi
Goldenweiser AA (1916) Culture and environment. Am J Sociol 21(5):628–633
Goldenweiser AA (1924) Race and culture in the modern world. J Soc Forces 3(1):127–136
GoUP [Government of Uttar Pradesh] (1967) Uttar Pradesh gazetteer – Pithoragarh district. Government of Uttar Pradesh, Lucknow
Hacking I (1995) The looping effects of human kinds. In: Sperber D, Premack D, Premack A (eds) Causal cognition: a multidisciplinary debate. Clarendon, Oxford, pp 351–394
Hancock M (2001) Home science and the nationalization of domesticity in colonial India. Mod Asian Stud 35(4):871–903
Handa OC (2002) History of Uttaranchal. Indus Publishing Company, New Delhi
Herod A (2011) Scale. Routledge, London
Howitt R (1998) Scale as relation: musical metaphors of geographical scale. Area 30(1):49–58
Husain Z (1995) Uttarakhand movement: the politics of identity and frustration. Prakash Book Depot, Bareilly
IWI [Illustrated Weekly of India] (1974) Bhotias of Kumaon (by Prakash Krishen), June
Jaffrelot C (2003) India's silent revolution: the rise of the lower castes in north India. Hurst, London
Jain NK (2009) Folk imagination: women in Indian love legends. Intersections: Gend Sex Asia Pac 22:n.p
Jenkins R (2008) Rethinking ethnicity, 2nd edn. Sage, London
Jones JP III, Woodward K, Marston SA (2007) Situating flatness. Trans Inst Br Geogr 32(2):264–276

Joshi PC (1986) Founders of the Lucknow school and their legacy: Radhakamal Mukerjee and D.P. Mukerji. Some reflections. Econ Pol Wkly 21(33):1455–1469

Joshi PC (2001) In the lap of the Himalaya: Gandhi's visit to Uttarakhand. Econ Pol Wkly 36(34):3300–3310

Kaiser R, Nikiforova N (2008) The performativity of scale: the social construction of scale effects in Narva, Estonia. Environ Plan D Soc Space 26(3):537–562

Kapila K (2008) The measure of a tribe: the cultural politics of constitutional reclassification in north India. J R Anthropol Inst 14(1):117–134

Karlsson BG (2003) Anthropology and the 'indigenous slot': claims to debates about indigenous peoples' status in India. Crit Anthropol 23(4):403–423

Karlsson BG, Subba T (eds) (2006) Indigeneity in India. Kegan Paul, London

Kaviraj S (2000) Modernity and politics in India. Daedalus 129(1):137–162

KBPF [Kumaon Bhotiya Peoples' Federation] (1947) The report of the Kumaon Bhotiya Peoples' Federation. Submitted before the minority sub-committee of the constituent assembly's advisory committee on 17th April 1947

Klenk RM (2004) 'Who is the developed woman?': women as a category of development discourse, Kumaon, India. Dev Chang 35(1):57–78

Kreutzmann H (2006) People and mountains: perspectives on the human dimension of mountain development. Glob Environ Res 10(1):49–61

Kroeber AL, Kluckhohn C (1952) Culture: a critical review of concepts and definitions. Vintage Books, New York

Lall P (1942) Hindu customary law in Kumaon. Superintendent, Printing and Stationary, United Provinces, India, Allahabad

Landor AH (1899) In the forbidden land: an account of a journey in Tibet, capture by the Tibetan authorities, imprisonment, torture, and ultimate release. William Heinemann, London

Landor AH (1905) Tibet and Nepal: painted and described by A. Henry Savage Landor. A. and C. Black, London

Latour B (2005) Reassembling the social: an introduction to actor-network-theory. Oxford University Press, Oxford

Leder S (2003) Zu einer Zeit, als Bäume und Gräser noch sprechen konnten: Sozioökonomie, Kosmologie und Mythologie der Rang-Shauka im zentralindischen Himalaya. Peter Lang, Frankfurt am Main

Linkenbach A (2007) Forest futures: global representations and ground realities in the Himalayas. Seagull, London

MacKinnon D (2011) Reconstructing scale: towards a new scalar politics. Prog Hum Geogr 35(1):21–36

Mansfield B (2005) Beyond rescaling: reintegrating the 'national' as a dimension of scalar relations. Prog Hum Geogr 29(4):458–473

Marston S, Jones JP III, Woodward K (2005) Human geography without scale. Trans Inst Br Geogr 30(4):416–432

Mawdsley E (1996) Uttarakhand agitation and other backward classes. Econ Pol Wkly 31(4):205–210

Meissner K (1985) Malushahi and Rajula: a ballad from Kumaon (India) as sung by Gopi Das. Part I: Kumaoni text, translation and appendices. Otto Harrassowitz, Wiesbaden

Michaud J (2012) Hmong infrapolitics: a view from Vietnam. Ethn Racial Stud 35(11):1853–1873

Moore A (2008) Rethinking scale as a geographical category: from analysis to practice. Prog Hum Geogr 32(8):203–225

Nautiyal S, Rao KS, Maikhuri RK, Saxena KG (2003) Transhumant pastoralism in the Nanda Devi biosphere reserve, India. A case study in the buffer zone. Mt Res Dev 23(3):255–262

Nawa K (2000) Ethnic categories and their usages in Byans, far western Nepal. Eur Bull Himal Res 18:36–57

Nehru J (1973) The tribal folk. In: India Ministry of Broadcasting and Information (ed) The tribal people of India. Government of India Press, New Delhi, pp 1–7

Nehru J (2007 [1947]) Tryst with destiny. In: Batabyal R (ed) The penguin book of modern Indian speeches. 1877 to the present. Penguin Books India, New Delhi, pp 327–329

Neveling P, Wergin C (2009) Projects of scale-making: new perspectives for the anthropology of tourism. Etnográfica 13(2):315–342

Pangtey SS (1992) Madhya Himalaya ki Bhotia janjati. Takshila Prakashan, New Delhi

Pant SD (1935) The social economy of the Himalayas. Based on a survey in the Kumaon Himalayas. George Allen and Unwin, London

Patel S (1988) Construction and reconstruction of woman in Gandhi. Econ Pol Wkly 23(8):377–387

Peabody N (2001) Cents, sents, census: human inventories in late precolonial and early colonial India. Comp Stud Soc Hist 43(4):819–850

Pels PJ, Salemink O (eds) (1999) Colonial subjects: essays in the practical history of anthropology. University of Michigan Press, Ann Arbor

Pfeffer G, Behera DK (2002) Introduction. In: Pfeffer G, Behera DK (eds) Contemporary society: tribal studies, vol 5, Concept of tribal society. Concept Publishing Company, New Delhi, pp 9–30

Pratt ML (1991) Arts of the contact zone. Profession 91:33–40

Purcell M, Brown JC (2005) Against the local trap: scale and the study of environment and development. Prog Dev Stud 5(4):279–297

Rangan H (2004) Of myths and movements: rewriting chipko into Himalayan history. Oxford University Press, New Delhi

Rawat DS, Sharma S (1997) The development of a road network and its impact on the growth of infrastructure. A study of Almora district in the central Himalaya. Mt Res Dev 17:117–126

Raypa RS (1974) Shauka: simavarti janjati. Raypa Brothers, Dharchula

RoC [Report of the Commissioner] (1952) Report of the commissioner for scheduled castes and scheduled tribes, vol 1. Government of India Press, New Delhi

Rudolph L, Rudolph S (2002) Living with multiculturalism: universalism and particularism in an Indian historical context. In: Shweder R, Minow M, Markus H (eds) Engaging cultural differences: the multicultural challenge in liberal democracies. Russell Sage, New York, pp 43–62

Rungmung (2013) Rung Kalyan Sangstha. Community Website. http://rungmung.net/index.php?option=com_content&view=article&id=1601:about&catid=88888940:about&directory=111. Accessed 20 June 2013

Rycroft DJ, Dasgupta S (2011) Indigenous pasts and the politics of belonging. In: Rycroft DJ, Dasgupta S (eds) The politics of belonging in India: becoming adivasi. Routledge, London, pp 1–14

Sahlins M (1993) Goodby to tristes tropes: ethnography in the context of modern world history. J Mod Hist 65(1):1–25

Sahlins M (1999) What is anthropological enlightenment? Some lessons of twentieth century. Annu Rev Anthropol 28:i–xxiii

Sax WS (1991) Mountain goddess: gender and politics in a Himalayan pilgrimage. Oxford University Press, New York

Sax WS (2009) God of justice: ritual healing and social justice in the central Himalayas. Oxford University Press, New York

Sax WS (2011) Religion, rituals, and symbols of belonging: the case of Uttarakhand. In: Pfaff-Czarnecka J, Toffin G (eds) The politics of belonging in the Himalayas: local attachments and boundary dynamics. Sage, Los Angeles, pp 167–181

Schlee G (2011) Afterword: an ethnographic view of size, scale, and locality. In: Schiller NG, Çağlar A (eds) Locating migration: rescaling cities and migrants. Cornell University Press, Ithaca

SCI [Supreme Court of India] (2000) State of Maharashtra vs. Milind & Ors, November 28

Shah A, Shneiderman S (2013) The practices, policies, and politics of transforming inequality in South Asia: ethnogrtaphies of affirmative action. Focaal J Glob Hist Anthropol 65(1):3–12
Sherring CA (1974 [1916]) Western Tibet and the British borderland. The sacred country of Hindus and Buddhists. With an account of the government, religion and customs of its peoples. Cosmo Publications, Delhi
Sinha S (2004) Colonial anthropology vs. indological sociology: Elwin and Ghurye on tribal policy in India. In: Subba TB, Som S (eds) Between ethnography and fiction: Verrier Elwin and the tribal question in India. Orient Longman, Hyderabad, pp 86–109
Sipal DR (n.d.) Chira! Darma ci rani. In: Son Mujir – Dugtu Sajir 'chira Darma ci rani'. N.p.
Smith N (1993) Homeless/global: scaling places. In: Bird J, Curtis B, Putnam T, Robertson G, Tickner L (eds) Mapping the future: local cultures, global change. Routledge, London, pp 87–119
Smith N (1995) Remaking scale: competition and cooperation in prenational and postnational Europe. In: Eskelinen H, Snickars F (eds) Competitive European peripheries. Springer, Heidelberg, pp 59–74
Smith N (2004) Scale bending and the fate of the national. In: Sheppard E, McMaster RB (eds) Scale and geographic inquiry: nature, society and method. Blackwell, Oxford, pp 192–212
Spurr D (1993) The rhetoric of empire: colonial discourse in journalism, travel writing, and imperial administration. Duke University Press, Durham
Srinivas MN (1956) A note on Sanskritization and Westernization. Far East Q 15(4):481–496
Srinivasan S (2005) Daughters or dowries? The changing nature of dowry practices in South India. World Dev 33(4):593–615
Srivastava RP (1953) 'Rang-Bang' in the changing Bhotia life. East Anthropol 6(3–4):190–203
Srivastava RP (1957) Marriage and divorce among the eastern Bhotias. Anthropol: Bi-Ann J Anthropol 4(1–2):34–43
Srivastava RP (1966) Tribe-caste mobility in India and the case of Kumaon Bhotias. In: von Fürer-Haimendorf C (ed) Caste and kin in Nepal, India and Ceylon. Asia Publishing House, Bombay, pp 161–212
Statesman (1981) The inner line legacy. Trend towards isolation, December 28
Stoler AL (1997) Carnal knowledge and imperial power: gender, race, and morality in colonial Asia. In: Lancaster RN, di Leonardo M (eds) The gender sexuality reader: culture, history, political economy. Routledge, New York, pp 13–36
Strachey H (1853) Physical geography of western Tibet. J R Geogr Soc Lond 23:1–69
Sundar N, Deshpande S, Uberoi P (2000) Indian anthropology and sociology: towards a history. Econ Pol Wkly 35(34):1998–2002
Swyngedouw E (1997a) Neither global nor local: 'glocalization' and the politics of scale. In: Cox KR (ed) Spaces of globalization: reasserting the power of the local. Guilford Press, New York, pp 137–166
Swyngedouw E (1997b) Excluding the other: the production of scale and scaled politics. In: Lee R, Wills J (eds) Geographies of economies. Arnold, London, pp 167–176
Tambiah SJ (1973) Dowry and bridewealth and the property rights of women in South Asia. In: Tambiah SJ, Goody J (eds) Cambridge papers in social anthropology, vol 7. Cambridge University Press, Cambridge, pp 59–169
Taylor C (1994) The politics of recognition. In: Gutman A (ed) Multiculturalism: examining the politics of recognition. Princeton University Press, Princeton, pp 25–73
Tillin L (2011) Questioning borders: social movements, political parties and the creation of new states in India. Pac Aff 84(1):67–87
Tiwari SC (1952) The distribution of blood groups among Bhotias of Almora district (U.P.). Man India 32:148–151
Tolia RS (2010) Great tribal diversity of Uttarakhand. Amtikar [magazine distributed by the Rang welfare association, Pithoragarh]:49–56

Traill GW (1992 [1832]) Statistical report on the Bhotiya mehals of Kumaon. In: Joshi MP, Fanger AC, Brown CW (eds) Himalaya: past and present, vol II. Shree Almora Book Depot, Almora, pp 99–154

Tribune (2012) Bahaguna ignored Dharchula constituency, says MLA, May 4

Tylor EB (1871) Primitive culture: researches into the development of mythology, philosophy, religion, language, art and custom. John Murray, London

UGP [Uttarakhand Government Portal] (2013) Tourist arrivals (Indian and foreign), 2010. http://uk.gov.in/pages/view/441-tourism. Accessed 13 Oct 2013

UHC [Uttaranchal/ Uttarakhand High Court] (2004) Bhotiya janjati sangharsh vs. Dr. R.S. Tolia. Judgement

UTDP [Uttarakhand Tourism Development Plan] (2008) Uttarakhand tourism development master plan 2007–2022. Final report. Government of India, Government of Uttarakhand, United Nations Development Programme, World Tourism Organization. http://uttarakhandtourism.gov.in/files/Volume%202%20%28low%20res%29.pdf. Accessed 13 Oct 2013

van Schendel W (2005) The Bengal borderland: beyond state and nation in South Asia. Anthem Press, London

van Schendel W (2011) The dangers of belonging: tribes, indigenous peoples and homelands in South Asia. In: Rycroft DJ, Dasgupta S (eds) The politics of belonging in India: becoming adivasi. Routledge, London, pp 19–43

van Schendel W, Bal E (2002) Beyond the 'tribal' mind-set: studying non-Bengali peoples in Bangladesh and west-Bengal. In: Pfeffer G, Behera DK (eds) Contemporary society: tribal studies, vol 5, Concept of tribal society. Concept Publishing Company, New Delhi, pp 121–138

Walton HG (1911) Almora: a gazetteer, vol XXXV of the district gazetteers of the united provinces of Agra and Oudh. Government Press United Provinces, Allahabad

Wimmer A (2013) Ethnic boundary making: institutions, power, networks. Oxford University Press, Oxford

Xaxa V (1999) Transformation of tribes in India: terms of discourse. Econ Pol Wkly 34(24):1519–1524

Xaxa V (2004) Women and gender in the study of tribes in India. Indian J Gend Stud 11(3):345–367

Chapter 4
Facets of Pastoral Mobility in a Market-Oriented Border Region

Abstract In this chapter I analyze how one subgroup of the Bhotiyas, the Darmani, practice mobility in the market-oriented border region of today's Kumaon Himalaya. The chapter is divided into three major parts. In the first part I scrutinize how recent policy initiatives have influenced the manner in which the residents of the Darma valley use different ecological belts through seasonal migration. While mobile animal husbandry is increasingly recognized as a viable livelihood strategy, very little is known about the socio-cultural meanings, motivations and forms of agency that underpin the mobile practices of mountain pastoralists. In order to gain a deeper insight into these embodied facets of pastoral mobility, the second part analyzes a ritual sequence that unfolds between the time of sowing and harvest in the seasonally inhabited summer settlements. These rituals are intimately tied to a wide range of agro-pastoral work activities and visibly connected through the establishment, renewal and removal of sacred tree trunks, called *alams*. However, it is not only forest but also livestock resources that figure prominently in acts of ceremony. The role of the yak, as illustrated in the third part, once stood at the centre of the traditional mortuary ceremony, called *gvan*, where it acted as a fellow wayfarer of the departed soul on its way to the land of the dead. Whilst the whole ceremony has undergone significant change since colonial times and was eventually renamed as *sarat*, one key element endured: the recitation of an elaborated oral text that is performed to guide the newly departed soul on its journey into the beyond. The chapter shows how mobility-related rituals foster people's responsiveness to the fluctuations of their dwelt-in environments: the seasonal cycle of weather conditions, the vegetative growth cycles of plants, and the life cycles of fellow human and other animal creatures. With a focus on the kinaesthetic and performative-poetic dimensions of this responsiveness, the study foregrounds the importance of bringing qualitative questions of meaning, value and purpose to bear in research on the numerous facets of mountain pastoralists' mobility.

Keywords Mobility • Mountain pastoralism • Dwelling • Ritual • Oral text • Kinaesthetics • Performance • Poetics • Bhotiyas • Darma valley • Kumaon Himalaya • India

C. Bergmann, *The Himalayan Border Region*, Advances in Asian Human-Environmental Research, DOI 10.1007/978-3-319-29707-1_4

Mobility is a prominent theme in high mountain research, especially in the analyses of pastoral communities' ways of life. Without doubt, the rhythms of movement in the Himalaya are broadly driven by seasonal variations, such as the duration of snow cover or the onset of the vegetation periods in different altitudinal belts (Bergmann et al. 2011). Classic accounts on this subject emphasized cross-regional similarities in vertical land use and specified mobility patterns as adaptations that reduce the risk of crop and livestock production under the harsh environmental conditions of mountain regions (Rhoades and Thompson 1975). Scholars frequently focused attention on the interrelationship of two main parameters, namely the size of livestock populations and the quantity of available resources.[1] Within the framework of such a quantitative metric, pastoralists predominantly figured as "politically passive migrants" (Agrawal and Saberwal 2004: 38), because their "decision making power and culture" was largely neglected (Kreutzmann 2001: 132; cf. Kreutzmann 2006: 51).

Negative perceptions of pastoral mobility gained a strong currency in the Theory of Himalayan Environmental Degradation (THED), a dominant narrative for that region during the 1970s and 1980s (Ives and Messerli 1989; Metz 2010). It postulated a direct relation between overpopulation, land degradation and deforestation, predicting severe soil erosion in the hills, and devastating floods in the Indo-Gangetic plain (Eckholm 1976). While these prognostic scenarios and causal explanations still inform policy-making in India and China today, they have meanwhile been challenged by a number of studies emphasizing the value of 'traditional' knowledge and postulating far more complex pictures of environmental change (Blaikie and Muldavin 2004). Stimulated by scholarship on the semi-arid rangelands of Africa (Scoones 1995; Niamir-Fuller 1999), a new scientific agenda for the assessment of mobile land-use in the Himalaya gradually emerged in which the proactive character of pastoralists' livelihood strategies is foregrounded (Nüsser 1998; Saberwal 1999; Kreutzmann 2012).[2]

[1] The concept of 'carrying capacity' is used to determine the number of animals that could be supported in any given area. The approach follows the so-called Lotka-Volterra model of predator/prey relations, which is summarized by Ingold (1980: 38) as follows: "Two species, one of which feeds on the other, must undergo perpetual and undamped oscillations in numbers. The predators usually increase to a point of overloading their food supply, then decline again through malnutrition, giving the prey a chance to increase, consequently allowing the predator to increase, causing the prey to decline and so on [...]". For calculating carrying capacity livestock is generally put in the position of the predator and fodder plants in the position of prey (Hjort 1982: 12–16).

[2] According to Scoones (1999: 490) this paradigm, which is also called 'new ecology', revolves around three major themes: "First is the concern with spatial and temporal dynamics developed in detailed and situated analyses of 'people in places', using, in particular, historical analysis as a way of explaining environmental change across time and space. Second is the growing understanding of environment as both the product of and the setting for human interactions, which link dynamic structural analyses of environmental processes with an appreciation of human agency in environmental transformation, as part of a structuration approach. Third is the appreciation of complexity and uncertainty in social-ecological systems and, with this, the recognition of that prediction, management, and control are unlikely, if not impossible".

Scholars have pointed out that the migratory routes of mountain pastoralists in many parts of the world are highly variable over time (Montero et al. 2009). Such transformations are not only linked to changing environmental conditions but also to broader socio-political processes that enable or constrain specific movements. The Sino-Indian border dispute, for instance, has exerted a strong impact on the management of pasture resources because it restricted certain patterns of movement and stimulated new policy initiatives toward mobile animal husbandry (Gerwin and Bergmann 2012). Such power dynamics also interact with and influence the discursive representations through which pastoralists' mobility is recognized in the public gaze (Gooch 2009). A shepherd has to constantly negotiate such representations in his daily encounters with government officials or resident user groups, for example when paying grazing fees or seeking permission to enter a forest. Whilst dealing with shifting environmental conditions, market relations, state policies and public discourses he will in general align his livelihood activities on the basis of numerous improvisational skills "that permit continuous flow in human performance of all kinds" (Richards 1993: 63). Primarily, these skills consist of pastoralists' embodied knowledge, which is generally acquired and activated in settings of practical activity and engagement – when doing or making things – as well as in the interpretation of experience (Bourdieu 1977; Ingold 2000: 157–171).

In Kumaon, the effects of government interventions into local strategies of natural resource use have received some attention, though most scholars restricted their perspective to the genealogy of a scientifically grounded forestry in the Lesser Himalaya and largely ignored pastoral groups like the Bhotiyas (Guha 1989, 2001; Agrawal 2001, 2005a, b). Guha exposes how early punitive guidelines for environmental conservation under British colonial rule interfered heavily with local interests and triggered long-term resistance. Agrawal further analyses the emergence of state-community partnerships, which culminated in formally recognized forest councils (*van panchayats*) during the 1930s. The *van panchayats* were officially recognized by the colonial state after a famous and often-described forest rebellion (Guha 2001). This makes them one of the oldest surviving examples of formally approved agreements for the governance of communal resources between state authorities and local user groups in the world. Around Munsiari and Dharchula most of these councils were introduced only after the border war with China in 1962, when the Indian state sought to enhance its presence in the Himalayan periphery (Nüsser and Gerwin 2008).

During the 1990s the Joint Forest Management Program, a pan-Indian scheme of community-based resource regulations that was co-financed by the World Bank, had a major impact on the *van panchayats*. Focusing on issues like participation, empowerment, and gender equality, an internationally recognized set of communal resource regulations was gradually standardized (GoU 2001).[3] Based on a Foucauldian framework of 'governmentality' Agrawal argues that these decentralized and participatory management procedures replaced repressive force with

[3] For an overview of the key policy issues in participatory forest management, including its implementation in the field, see Springate-Baginski and Blaikie (2007).

engaging and educative mechanisms of control and compliance.[4] In these councils rules are implemented, fine-tuned and propagated on the ground of disciplinary practices through which "those subject to government realize the importance of careful use of environmental resources such as forests and pastures or clean air and water" (Agrawal 2005b: 187). In other words, the functioning of the *van panchayats* is driven by what Foucault (1988) has termed as 'technologies of self and power', i.e. a set of discursive practices and strategies through which actors internalize specific values that are bound to the powers of 'expertise' and linked to the interests and decisions of the state.

According to Agrawal (2005b: 178–179) forest councils' mode of operation is therefore best understood as a form of 'intimate government': state officials devolve authority to communities and facilitate localized initiatives through the creation of training opportunities and the allocation of funds. By means of these measures villagers can then decide as to which conservation principles they would like to apply in a forest. The apparent effectiveness of these decentralized means of power indicate, as Agrawal argues, that the dichotomy between 'traditional', or 'indigenous', and 'scientific', or 'expert', knowledge is in many ways problematic and untenable since each of them is inherently heterogeneous and contested (Agrawal 2005c). Agrawal, however, still considers the environment as a non-human domain, or as a framework of *external* physical constraints and potentials that awaits *rational* management (Skaria 2005). Following from this, environmental knowledge is assessed here in terms of an assumed trans-cultural rationality that most closely matches governments' understanding of environmental conservation, resource efficiency and sustainability.

While Agrawal admits that local cultural practices also shape the manner in which communities engage with their surroundings, he nevertheless proceeds with the assumption that these do not, or at least not in any significant way, affect the long-term success, or failure, of implemented programs for environmental protection and sustainable development. Nevertheless, the manner in which people use and care for their environs is almost always entwined with numerous practices, values and commitments that will never be touched upon in the officially sanctioned guidelines of a *van panchayat* (Nightingale 2005). Already Leach et al. (1999: 240) have pointed out that

> multiple [social] institutions are involved in natural resource management. Most of these are not dedicated to the purpose of or dependent on it in any functional way – marriage and kinship exchange networks 'do' many other things besides their role in land access for instance – yet are important in mediating the endowments and entitlements of certain social actors.

[4]Foucault (1991, cf. 1977) famously phrased the notion of 'governmentality' as the 'conduct of conduct' in order to describe the manner in which modern governments exert power by disciplining its subjects through the distribution of politico-economic resources and the spread of 'expert' knowledge. This happens particularly through officially sanctioned institutions such as schools, hospitals, prisons, or, as in the present case, forest councils. By participating in the governing of the environment through such state-community partnerships "new beliefs and, thus, new subjectivities" (Singh 2013: 190) are created. To describe the technologies and procedures that are at work in the creation of these subjectivities, Agrawal (2005a) coined the term 'environmentality'.

It is on such grounds that Ingold and Kurttila (2000) have delivered yet another answer to the question of why the modernist distinction between 'indigenous' and 'scientific' environmental knowledge is misleading. They start with the assumption that any kind of environmental knowledge is engendered by 'practices of locality'. A key difference between the knowledge of a remote sensing specialist and, let's say, a mountain shepherd lies neither in its epistemological status nor in its higher or lower level of heterogeneity; it rather concerns the nature of the skills through which these actors perceive and deal with such phenomena as changes in the tree and vegetation species composition, grassland productivity and ecosystem health "*by way of* their practice" (Ingold 2011: 159; cf. Robbins 2001).

Through their respective skills both the shepherd and the remote sensing specialist learn something about the very same thing, 'the environment'. The issue is, however, that these skills are most likely guided by different 'modes of apprehension' (Ingold 2000: 323). In a nutshell: the former apprehends the environment by tuning-in with its rhythms and components; the latter does the same by withdrawing from any active participation with it. One sees a world that is 'inhabited', the other a world that is 'occupied' (Ingold 2011: 120). It is important to note, however, that these modes of apprehension are not mutually exclusive for Ingold (2000: 216) but rather "caught up in the dialectical interplay between engagement and detachment, between human beings' involvement in the world and their separation from it". Yet, modern science and technology often makes it appear as if the engaged-mode of apprehension is only of secondary importance.

The remote sensing specialist might also fall in with the dwelt-in environment, for instance when generating thematic maps based on land use and cover classifications with respect to vegetation units and agricultural areas. This work requires a constant combination of engaged fieldwork and detached digital image processing to pre-define, define, and cross-check ('ground truth') the classes and the accuracy of classification results. Moreover, one could even argue that working in the remote sensing lab does itself involve forms of engagement with the constituents of a working environment such as computers, algorithms and colleagues. By the same token the shepherd might partially withdraw from his active engagement with the environment, for instance when adjusting his movements through the usage of mobile service communications to access weather forecasts or when performing a ritual that moves his attention away from the rhythms and relations of everyday life.

Ingold analyzes such diverse forms of human-environment interaction in terms of what he calls the 'dwelling-perspective'. The term 'dwelling' gained currency in phenomenological thinking especially through Heidegger's (1971: 141–160) famous essay 'Building, Dwelling, Thinking', to which Ingold (2000: Chap. 10) explicitly refers. Heidegger focuses upon two major questions, namely 'What is it to dwell?' and 'How does building belong to dwelling?' He answers these questions by rejecting the means-end relationship according to which humans attain to dwelling by means of building. According to Heidegger one cannot assume that buildings guarantee dwelling. In other words, there is a difference between the physical structure, a house for instance, and the home, or the site where dwelling occurs. The two modes of building, viz. cultivating and constructing, are rather themselves already acts of dwelling: “We do not dwell because we have built, but we build because we dwell, that is, because we are dwellers” (Heidegger 1971: 148). Ingold adopts the term ‘dwelling’ to scrutinize the manifold ways in which actors become actively involved with and get to know their environs through their hands-on experience.

In terms of dwelling actors acquire environmental knowledge by incorporating the features of the physical landscape into a rhythmic pattern of everyday activity, which Ingold calls ‘taskscape’ (Ingold 2000: 194–198). Reduced to a simple formula, the taskscape is the landscape put into practice and the landscape is the material record of the activities that comprise the taskscape. For the skilled practitioner both tend to become indistinguishable in that she can hardly separate her activities from the world in which they are performed (Mazullo and Ingold 2008: 37). Ingold uses the adaptive interaction between landscape and taskscape to dismantle a further dichotomy that will be of crucial importance to the remainder of this chapter: the widespread distinction between “people’s practical-technical interaction with environmental resources in the context of subsistence routines and their mytho-religious or cosmological construction of the environment in the context of ritual and ceremony” (Ingold 2000: 56). It is easy to recognize that the distinction rests upon the Cartesian dualism between the material and the mental, which appears here in the guise of human actors who act upon the natural environment versus human actors who construct the same according to some culturally given categories.

For Ingold, however, both sets of activities describe, in the first place, ways of dwelling whereby both rest upon the mutually constitutive interrelation between persons and their environs. Let’s stay with the example of the mountain shepherd to elucidate this point. Whilst realizing a repeated transition from one pasture area to another he will attend to the unstable concentrations of biotic resources. In other words, when moving within or between different ecozones he will not simply bring his flock to where the pasture is but proactively pick up potential clues in the ever-shifting textures of his surroundings. In that way he enhances the quality of nutrients available to his animals (Nüsser 1998: Chap. 4; Krätli and Schareika 2010). So what is the shepherd doing when performing a ritual or when reciting a myth? Is he then shifting his attention from the ‘real’ world of material dynamics and energy flow to a mere metaphorical or mental representation of that world?

According to Ingold such an appraisal would be rather delusive.[5] In acts of ceremony people are similarly attentive to and engaged with the spatial and temporal variability of their ‘real’ surroundings. Previous studies on mountain pastoralism have already described rituals as important social institutions that allow people “to process information concerning the distribution of resources, the scheduling of agricultural tasks, and the coordination of conflicting demands on labor” (Guillet 1983: 566). However, these practices were basically considered as instrumentally rational

[5] In her classic work *Purity and Danger*, Mary Douglas (1966: 148) similarly remarked that ‘primitive cultures’ do not use bodily magic as an escape from reality but rather “to confront experience with its inevitable pains and losses”.

'control mechanisms' or 'communicative devices' that help people to reduce the risk of livelihood failure in a hazardous environment. This environment, however, was considered as being external to society. Without denying that rituals play a role in relation to mountain pastoralists' handling of risks and vulnerabilities, I argue that these practices – like all processes of dwelling – do not only allow people to act *upon* but also to act *within* an environment that is as much a biophysical as a socio-cultural phenomenon (Taylor 2015: 23–25).[6] The dynamic and contending forces that shape such environments are, for instance, seasonal weather conditions, the vegetative growth cycles of plants, the (livelihood) activities of fellow human and other animal creatures as well as the perceived needs of ancestral beings, deities and spirits.

Building on these introductory deliberations on Ingold's dwelling-perspective, I analyze in this chapter how the Bhotiyas practice mobility in the market-oriented border region of today's Kumaon Himalaya. The chapter, which places particular emphasis on the residents of the Darma valley, is divided into three major parts. In the first part I scrutinize how recent policy initiatives have influenced the manner in which the Bhotiyas use different ecological belts through seasonal migration. While mobile animal husbandry is increasingly recognized as a viable livelihood strategy, very little is known about the socio-cultural meanings, motivations and forms of agency that underpin the mobile practices of mountain pastoralists. In order to gain a deeper insight into these embodied facets of pastoral mobility, the second part explores a ritual sequence that unfolds between the time of sowing and harvest in the seasonally inhabited summer settlements of the Darmani.

These rituals are intimately tied to a wide range of agro-pastoral work activities and visibly connected through the establishment, renewal and removal of sacred tree trunks, called *alams*. However, it is not only forest but also livestock resources that figure prominently in acts of ceremony. The role of the yak, as illustrated in the third part, once stood at the centre of the traditional death ceremony, called *gvan*, where it acted as a fellow wayfarer of the departed soul on its way to the land of the dead. Already under British colonial rule the mortuary rites fuelled numerous discussions, especially because the yak was ultimately killed during the ritual. Whilst the whole ceremony has undergone significant change since then and was eventually renamed as *sarat*, one key element endured: the recitation of an elaborated oral text, which is performed to take the departed soul on a long and dangerous journey into the beyond. The overall analysis shows how mobility-related rituals foster people's responsiveness to the fluctuations of their dwelt-in environments. By foregrounding the kinaesthetic and performative-poetic dimensions of this responsiveness, the study brings qualitative questions of meaning, value and purpose to bear in research on the numerous facets of mountain pastoralists' mobility.

---

[6] It is very common in North India that people refer to those acts that scholars label 'ritual' as 'work' (*kam*) (Sax 2010: 4).

4.1 The Bhotiyas' Seasonal Migration in the Context of Emergent Policy Initiatives

Between 1961 and 2001 the total number of people who are practicing seasonal migration in the Bhotiya valleys of Kumaon has more than halved.[7] The negative trend is clearly visible in Milam, the uppermost village of the Johar valley. In this village the fall in population has been the most drastically, namely from 600 households in the 1930s (Pant 1935: 240) to 300 in 1950 (Murray 1951) to then only 18 in 2004 (Nüsser 2006: 20). Although the figures are increasing again in some villages, the former winter settlements at the lower margins of the High Himalaya have clearly become the economic hub and centre of life for the majority of people. According to Kreutzmann (2012: 8) it can be observed in many localities of High Asia that the permanently settled operational base has grown in importance for (agro-) pastoral communities, while the contribution of mobile animal husbandry for securing a livelihood decreases.

Well-off community members, often with the help of urban-dwelling relatives, acquired property and built modern residences in these well-accessible parts of the valleys. Such property constitutes an important future investment, especially as an old-age provision.[8] Furthermore, a number of new 'colonies' have emerged, where people maintain larger plots of agricultural land. Examples of such colonies are Sobla (with families from Bon, Baling, Son, Nagling), Khet (with families from Baling) and Jamku (with families from Son and Nagling) (Fig. 4.1).[9] Most of the land was purchased after the Sino-Indian border war in 1962, when these families decided to compensate for the discontinuation of trans-Himalayan trade by intensified and all-year crop farming. The discontinuation of trans-Himalayan trade was in general accompanied by a decrease in sheep and goats, and an increase in the number of cattle (GoU 2003). The latter are the preferred animals of sedentary agriculturalists. Such new trends in livestock production are causally related to various environmental transformations (Gerwin and Bergmann 2012: 100).

Around Munsiari, for instance, the herbaceous understorey of montane oak forests became progressively exploited for pasturing cattle, supplemented by stall-feeding during the winter months. The reduction in mobile animal husbandry further led to the cessation of regular intentional burnings to sustain the grass cover, and facilitated an increase in multi-functional trees (mainly *Alnus nepalensis* and *Aesculus indica*) as well as in various fruit tree species around the villages that

---

[7] These trends are shown by the demographic data the author collected at the Directorate of Census Operations, U.P. Lucknow, India.

[8] Urban-dwelling community members undertake such investments also for emotional reasons, such as senses of place and belonging.

[9] Sobla was an important transit point and stock of merchandize for the Darmani traders, and it still is for contemporary shepherds and muleteers who are transporting goods, especially for the Indian Army and the Indo-Tibetan border police (ITBP). Up until now villagers graze their yaks and yak-crossbreeds in the vicinity of Sobla for a fee of INRs 25–35 (rate for 2009) per animal and month to the responsible *van panchayat*.

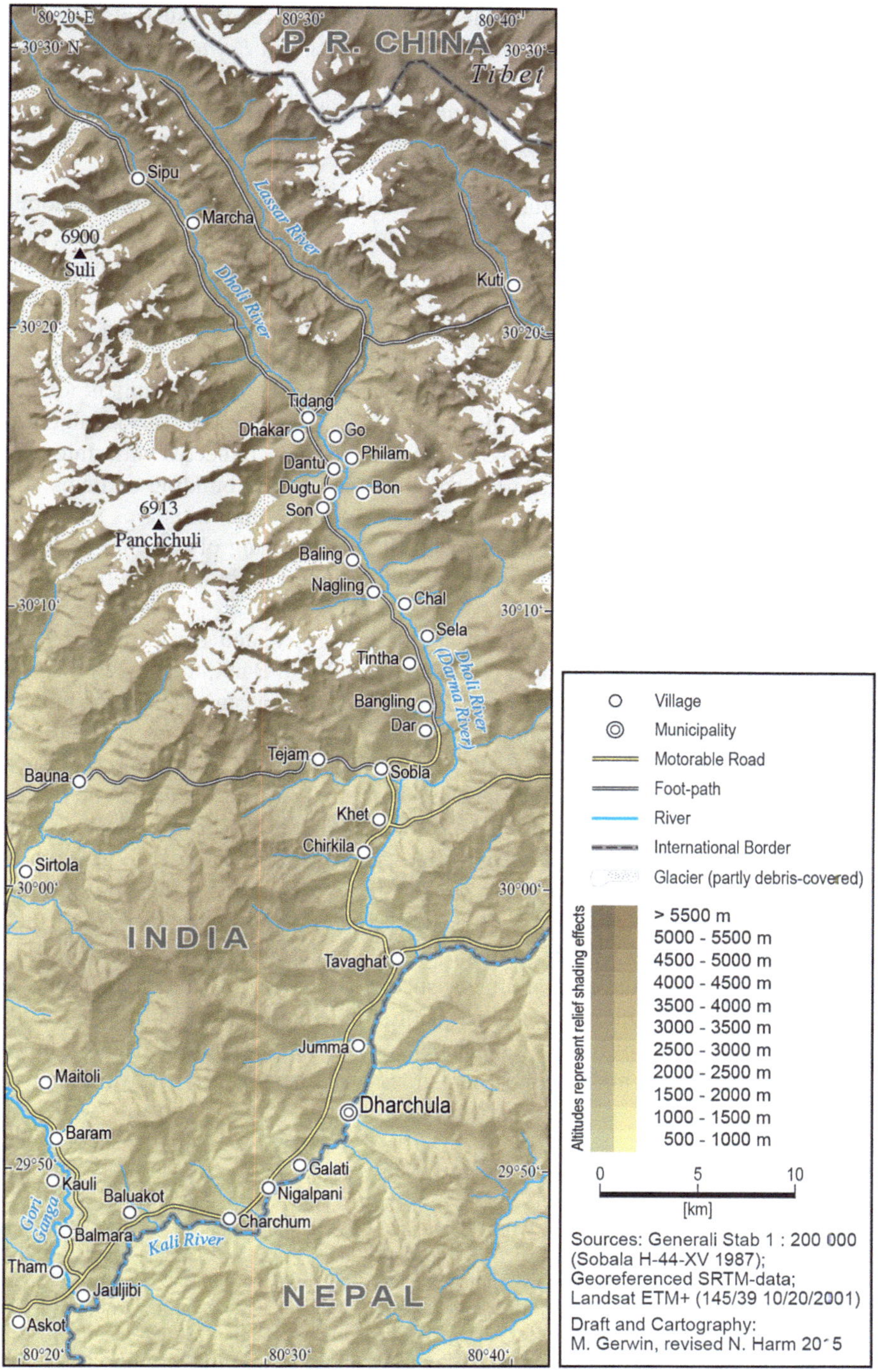

Fig. 4.1 The Darma valley

constitute this municipality. Many of the remaining areas are often strongly depleted and transformed into scrubland (with *Prinsepia utilis* and *Pyracantha crenulata*), and grasslands with less palatable grass species (e.g. *Cymbopogon munroi*, *Muhlenbergia duthieana* or *Themeda triandra*). Good fodder, particularly for cattle, has become scarce around Munsiari, fueling regular conflicts between resident communities.

However, it would be premature to conclude that the Bhotiyas' mountain mobility system represents a remnant of the past, doomed to vanish into oblivion. It is rather the case, as I shall argue, that the different altitudinal belts are combined and integrated in new ways. In a situation of federal competition for the allocation of funds, the Government of Uttarakhand set up a policy agenda aiming to align market-based approaches of development with mountain-specific requirements. This was backed by a more confident policy of India's central government toward its disputed boundary with China (Mohan 2007: 14–19). Through various flagship initiatives the valleys inhabited by the Bhotiyas have been connected with state- and nation-wide development programs. Examples include the building of dams and road infrastructure or the promotion of tourism. Whilst these initiatives have provided an important impetus for investment, the restructuring of regional development also involved targeted measures to directly support mobile animal husbandry (Gerwin and Bergmann 2012).[10]

By following a global scientific discourse, as propagated for instance by the Food and Agriculture Organization (FAO) and the World Bank, the Government of Uttarakhand has increasingly endorsed pastoral practices as an efficient instrument for poverty reduction and sustainable livelihoods (Tolia 2010). As a result, the 'tribal and transhumant strategy' (GoU 2004: 5, 2010a) of the Bhotiyas is promoted as an activity that needs to be strengthened, not least because it can also be turned into an asset for sedentary communities. A case in point is, for instance, when shepherds negotiate customary agreements with organic farmers in order to graze their animals on harvested fields in return for fertilizing manure and the sale of dairy products. A flagship initiative in this regard is the Decentralized Watershed Development Project (DWDP), which aims to align a participatory framework to the requirements of tribal communities, especially those who – like the Bhotiyas – have made a name for themselves in terms of mobile animal husbandry.[11]

Already at the time of India's independence the Bhotiyas had urged state administrators to grant "goat and sheep owners grazing rights in forests all over Kumaon

[10] A brief assessment of the region's tourism development has been provided in the previous chapter.

[11] The DWDP was co-initiated by the World Bank and is now largely financed via the Tribal Sub-Plan. It varies significantly from the expert-designed soil and water conservation as well as watershed management procedures that were implemented in the Himalaya with the rise of the THED in the 1980s (Blaikie and Muldavin 2004: 3). A major difference is that recent interventions leave a greater scope for herders' preferences and decision taking procedures. However, interviewed shepherds complained that the forested areas covered under the DWDP are not sufficient and that government employees would often still not take their voices and concerns serious enough.

particularly during winter months" as well as to provide "them suitable halting places (Parao) along their routes" (KBPF 1947: 19). Under the DWDP such demands are increasingly put into practice, often in collaboration with local *van panchayats*. The aim is to implement an action plan that is informed by what both settled and migratory communities experience as culturally compatible, economically feasible and environmentally sustainable (GoU 2004: 19). One fundamental shift in policy concerns the manner in which access to forests in the Lesser Himalaya is being regulated. Formerly, the forest division officers resorted to mere quantitative measures, such as carrying capacity, to specify the number of animals that were allowed to enter a specific forested area (GoU 2004: 14). The shepherds' stays in allotted grazing tracts of state forests were then often limited to 15–20 days (Nüsser 2006: 19).

Until present times the Bhotiya shepherds complain about various problems they face in the *Bhabar* during the winter months: from a lack of concern, corruption and fraud on the side of forest officials to mistreatment and animal thievery on the side of outsider groups. Some wealthy flock-owners have leased land in these regions to guarantee more suitable conditions or pursued their interests through the courts. In order to tackle the problem at its roots, the DWDP has promoted the recovery of communal regulations with regard to grazing. Through the targeted augmentation of fodder and biomass resources, the provisioning of better shelter facilities and the improvement of water availability in the respective forests, government officials try to reduce potential conflicts between settled and migratory populations. The by far greatest success, however, has been registered by various veterinary services and training establishments that have been launched along the shepherds' migratory routes to enhance animal healthcare and decrease animal mortality rates.[12]

While heavily reduced in number, sheep and goat rearing is still the largest and most profitable component of animal husbandry in the Bhotiya valleys. The majority of migrating households possess only a small number of these animals, and often none at all. Livestock owners typically merge their flocks until a suitable size is reached, and this *toli* (herd of sheep and goats) is then looked after by shepherds hired on contract.[13] A few non-migrating Bhotiyas also own flocks, some of a considerable size. They resort to similar arrangements whilst pursuing other occupations within and outside the agrarian sector. The contracted shepherds – both Bhotiyas and non-Bhotiyas – often also bring along their own animals. They mostly maintain strong, i.e. kinship-based or kinship-like, relations with residents of particular settlements, so that access to pasture grounds is negotiated and granted in informal ways (Photo 4.1).

---

[12]The veterinary camps provide basic but effective medical treatments such as de-worming. These services are offered free of charge to shepherds. A veterinary camp that is regularly frequented by the Darma shepherds in the spring is located near Champawat.

[13]The Darmani used to employ people from villages in the lower parts of their valley as shepherds. These people are referred to as 'Anwals', which is also a locally used term for 'shepherd' (Bisht 2001). Since the 1990s Gaddis from Himachal Pradesh increasingly take on this work (Axelby 2007; Wagner 2013). Some of them have intermarried with Bhotiya families.

Photo 4.1 Contracted shepherd in the Darma valley (Photograph © Christoph Bergmann, June 4, 2009)

In recent years a growing number of outsiders' flocks have begun to enter the high valleys (Bergmann et al. 2012: 183–184). The respective non-Bhotiya owners are from villages around Munsiari and Dharchula, the southern Pindari valley, Himachal Pradesh and far-western Nepal. These external user groups form a valuable income source for the *van panchayats* that now manage the alpine pastures surrounding the high altitude settlements. They levy a monthly grazing fee of Indian Rupees (INR) 3–5 per sheep or goat and up to INR 20 per horse (one USD equals approx. INR 45).[14] Most of these councils have only recently been (re-)established with the help of various non-governmental organizations (NGOs), which started to operate in the area, especially in the Johar valley, from the 1990s onwards. The rise of NGOs in Kumaon forms part of an Indian-wide trend to facilitate a decentralization of governance and development more efficiently and cost effectively (Townsend et al. 2004).[15] Both the traditional woolen-industries and the regional livestock sector have been transformed by distinct NGO activities.

While the Bhotiyas' woolen industry received little state support after the closure of the Indo-Tibetan border (Prasad 1989: 118–122), the increased presence of NGOs has had more of a positive impact in promoting local processing methods, procuring sufficient raw materials and marketing the finished products. Especially during the winter months, the processing of wool remains a primary occupation,

[14] All exchange rates were collected in 2009 if not otherwise mentioned.

[15] Whilst decentralized development activities seek to realize a greater integration of local users, they are still widely criticized by environmental activists and NGO-workers. One point of critique, for instance, is the increasing bureaucratization of organizational structures and annual implementation plans, which are seen as being far beyond the educational level of many local users (Sarin 2001).

particularly for women. The main manufactured products include carpets, blankets and clothes, which are sold at regional bazaars and at annual fairs for example in Jauljibi, Bageshwar and Nainital. In households that do not own any animals, these items are still produced by using purchased raw materials. Especially in Munsiari local NGOs arrange for a secure supply of high quality wool. In 2010 alone more than 6 tons of wool were transferred from Dharchula to Munsiari. But NGO-workers have also achieved significant developments in relation to the regional livestock sector, as the example of the re-introduction of yaks in a tributary valley of Johar, called Ralam, illustrates.

After the cessation of trans-Himalayan trade most of the yak herds declined until the last herd disappeared in the 1970s in the Johar valley (Bergmann et al. 2012: 184–185). A local NGO organized the revival of yak breeding in 1996 by importing a nucleus herd of 20 purebred yaks from Tibet. Johari herders support this endeavor to the present and promote inter-species breeding of yaks and cows for milk production and ploughing as well as for their use as pack animals during migration. The presence of these animals has the potential to influence an earlier than normal upward spring migration as people avoid grazing them at lower altitudes for too long (FES 2004: 19). In the Darma valley, yaks and their hybrid crossbreeds are still reared – in small numbers – even without NGO support. Some villages have established collective funds to cover the costs for keeping purebred male yaks at stud. Villagers purchase these animals in the adjoining region of Nepal, where a supply from Tibet is secure. Existing kinship ties to Nepalese traders facilitate these transactions.[16]

Another field of land use that has become extremely important since the late 1990s is the commercialized extraction and farming of medicinal and aromatic plants (Bergmann et al. 2012: 185–186). The Uttarakhand Government supports these endeavors, especially for the cultivation and merchandizing of *Allium stracheyi* (*jambu*) and *Carum carvi* (*thoya*).[17] The state-sponsored Herbal Research and Development Institute supervised the project in 2010, which has proved to be a great success in the Johar valley. At that time the initiative was in its testing phase in the Darma valley, where the collection of these and other plant species in the wild was still a more feasible option and widely practiced by the residents of the valley. Particularly worth mentioning are *Aconitum heterophyllum* (*atis*) and *Picrorhiza kurroa* (*kutki*), which are collected between August and November.[18] Of utmost interest to gatherers all over the region, however, is *Cordyceps sinensis* (locally known as *kida*, *bu* or *yarsha gumbo*), a fungus that parasitizes the caterpillar larvae of ghost moths (*Lepidoptera* spp.) in altitudes between 3,000 and 5,000 m (Winkler 2008).

---

[16]These animals are generally purchased in Tinkar and Changru. These high mountain settlements are situated across the Kali River in far-western Nepal and inhabited by the Byansi. They are accessible from the upper Darma valley via the Shinla pass.

[17]Both *jambu* and *thoya* are widely used as condiments.

[18]*Atis* is used as a remedy against fever and pain, *kutki* for digestive disorders and dysentery. Gatherers sell the roots to street vendors and pharmacists in Dharchula and Pithoragarh.

Photo 4.2 *Cordyceps sinensis* (Photograph © Christoph Bergmann, May 25, 2009)

This fungus is highly appreciated as a tonic, aphrodisiac and status symbol among the middle class in China, where large quantities of it are sold (Winkler 2010).[19] It is extensively gathered from May to July and then sold at highly fluctuating rates ranging between INR 200,000 and 700,000 (approx. USD 3,200–15,000) per kilogram to middlemen in the markets of Munsiari and Dharchula. The collection of this lucrative item widely contributes to the household income of both Bhotiya and non-Bhotiya residents, ranging from INR 15,000 (approx. USD 340) to more than INR 75,000 (approx. USD 1,685) in one season (Photo 4.2).[20] In the absence of a clear legal framework the selling of this item was treated as a criminal offence in Dharchula, which affected price formation and strengthened the expansion of unregulated (or informally regulated) market channels.

At the beginning of the boom in the early 2000s primary collectors used to sell the fungus to external brokers who visited the upper settlements at the end of June. However, due to reinforced roadside checks by state authorities, the items' sale gradually shifted to Dharchula and Munsiari. Some Bhotiyas have established themselves as national and international brokers who engage in this contraband trade with wholesale buyers from Delhi and Kathmandu. The economic value of these transactions can be estimated from the information given by one of these brokers. He reported that an agent from Kathmandu annually sends a helicopter to the Nepalese side of Dharchula (called Darchula) to purchase larger quantities of *C. sinensis* from various regional middlemen. However, most merchandize is

[19] An early reference regarding the trade of *C. sinensis* from Tibet to Chinese markets is given in Coales (1919: 244) who lists it under the name of 'chungtsao' or 'insect grass' that was "supposed to be an excellent restorative to weak constitutions".

[20] An average household engaged in the collection of *C. sinensis* in Son and Dugtu village (Darma valley) collected between 150 and 250 g in 2009–2010.

transported by car. Due to the large sums of money involved, the trade has given rise to mafia-like conditions, including protection rackets and armed robbery.[21]

These activities, which draw an increasing number of people to high-altitude grasslands, raise new opportunities and challenges for local residents and policy-makers. There have been numerous discussions in the Bhotiya villages about how to regulate and de-criminalize this new form of resource use. Most villages, for instance, adopted strict rules to enclose their grasslands and keep external groups out. Further, officials from the Forest Department began to interact more regularly with primary collectors with the aim of implementing specific procedures, such as the rotational opening of grasslands for *kida*-extraction to enhance sustainable harvesting. Only by September 2010 did the Uttarakhand Government authorize the *van panchayats* to officially intervene in this matter (GoU 2010b).[22] This happened under pressure from regional political leaders. Villagers are now obliged to sell their collected yields to *panchayat* officers who then resell them to external buyers. Five percent of the profits thus secured are charged as royalty. However, the unsupervised sale between primary collectors and regional brokers still continues, especially because many of the high altitude localities in which collection takes place are hardly or not at all kept under surveillance. At the same time the Bhotiya community launched a campaign to include the lucrative fungus in the list of officially approved Indo-Tibetan trade items across the Lipu Lekh pass, thus facilitating its direct sale to Chinese customers in Taklakot (Purang).

During the *C. sinensis* harvest period, the population in the upper valleys increases by approximately one-third. This 'gold rush' has resulted in new patterns of migration, where people join the summer settlements for shorter periods of time. However, the resulting temporary union of people has led to diverse forms of cooperation for planting and harvesting crops in the upper settlements. The main field crops include different varieties of buckwheat (*Fagopyrum esculentum* and *F. tataricum*), barley, and potato as well as legumes, which are all mostly cultivated for subsistence.[23] In recent years young families in particular have started to experiment with the cultivation of vegetables, such as cabbage, carrots and pumpkins. Once road access is provided, they hope to intensify these activities and sell part of their harvest to vendors in Dharchula and Pithoragarh.[24]

[21] In 2009, for instance, two young men from Dharchula were shot on returning from New Delhi after having sold a larger quantity of *C. sinensis* to international brokers.

[22] In 2009 buyers could legally purchase the fungus in Munsiari, while such commercial transactions remained illegal in Dharchula. While the new policy indicates a stronger presence of officially sanctioned schemes to regulate pasture use, these developments are not uniform. Without strong support from NGO workers and rather low efforts on the side of the Forest Department to re-work and control the extraction of forest and grassland resources, *van panchayats* remained less important in the Darma valley than in the neighboring Johar valley.

[23] In the Johar valley crop cultivation plays only a marginal role nowadays. Most households concentrate on medicinal plants and small-scale gardening. Medicinal plants, such as *Allium stracheyi* (*jambu*) and *Carum carvi* (*thoya*), are often cultivated in plots that are located in the ruins of old houses. Contrary to traditional village fields these plots are irrigated.

[24] Also the year-round presence of the Indo-Tibetan Border Police (ITBP) and the Indian military has brought in new customers for local field crops, vegetables and hay in the upper settlements.

To sum up the present situation: one can say that seasonal mobility across different altitudinal belts still is of significant economic importance, and that it has also come to be politically endorsed. After Uttarakhand's independence in the year 2000, the new government almost instantly promised that it would enhance the development of its remote tribal regions. For that purpose market-based approaches were combined with a mountain-specific policy. The Border Area Development Program and the Tribal Sub-Plan (TSP) are the main funding sources for this endeavor. Although the average cash income of migrating villagers is lower than that of their non-migrating counterparts, the prevalence of small-scale agriculture, the rising demand for medicinal and aromatic plants as well as cash remittances from urban dwelling relatives have resulted in an often acceptable level of subsistence. While the rearing of sheep and goats only 'pays' for those who own a larger number of animals, valley dwellers still profit indirectly from the increase in outsiders' herds that enter their high valleys through the fees imposed by the *van panchayats* for access to pastures. In the midst of these changes a renewed increase in seasonal migration toward the upper settlements is noticeable in recent years.

In what follows I ask whether and how ritual practices are integral to the Darmanis' overall mobility system. I argue that these acts of ceremony – like other work tasks – form part of people's ongoing engagement with their dwelt-in environment, including the multiple power relations inscribed therein. This is not to say that ritualized and quotidian activities are indistinguishable. On the contrary, rituals are effective because they are designed and orchestrated in ways that clearly differentiate them from everyday forms of activity, a process that Bell (1992: 88–93) describes as 'ritualization'. My point is, however, that rituals do not serve the purpose of structuring the environment according to some pre-given cosmological conception but rather to discover meaning (and perhaps also moral guidance) from 'clues' in the landscape that otherwise remain as background noise (Ingold 2000: 208). Such clues condense "otherwise disparate strands of experience into a unifying orientation which, in turn, opens up the world to perception of greater depth and clarity" (Ingold 2000: 22). Building on this observation I examine rituals as an embodied practice of pastoral mobility that takes shape in the relational contexts of the Darmanis' active involvement with each other and their environmental surroundings.

4.2 Rituals as an Embodied Practice of Pastoral Mobility

Preparations for the upward migration typically start at the end of February. Much needs to be done, including the purchasing of large quantities of food (e.g. rice and cooking oil) and other necessities at the Dharchula bazaar; making arrangements for school-aged children's stay with relatives in the lower settlements; obtaining the required Inner Line permits for the contracted shepherds from the sub-divisional magistrate; collecting the yaks and yak-crossbreeds from the winter pastures in the lower parts of the Darma valley; and transporting the packed baggage to the transit camps near Sobla.[25] The shepherds are amongst the first who then move toward the upper valley at the end of March. They advance slowly and eventually pitch a temporal campsite in the mid altitudes between the villages of Sela and Baling, where they wait for the residents of the settlement with whom they are connected to pass.[26]

The majority of migratory households only set off from mid-April onwards. On the day of their departure non-migrating neighbours and relatives come by to bid them farewell and wish a safe journey. Some *cyakti* (liquor) is ritually consumed whilst asking the deities for protection, before the migratory group begins to move. The daily distances covered are between 8 and 15 km, which are largely walked without a break to avoid the animals getting restless. These animals typically include some cattle, yak-cattle crossbreeds and, occasionally, also some horses.[27] The complete upward migration takes between 2 and 10 days, depending on weather conditions and the length of the overall route, as well as on the number and the age of migrating people and animals. The night camps are established either in the open or, on bad weather days, under plastic tarpaulins that are tightened over piles of luggage. The woolen blankets and carpets that are placed under the animals' packsaddles to prevent them from being injured by the heavy loads serve as a bivouac. Besides setting up the camp, people need to gather firewood and fodder grass, bring their animals to a water source, and prepare something to eat for themselves. Sleep periods are therefore rather short.

Since it takes about 2 h to get ready again in the morning, the migratory teams break camp at around 4 o'clock in the morning in readiness to move again at dawn's first light. The days of migration are thus rather strenuous and everybody is pleased when the upper settlement is finally reached. But also there the workload remains comparatively high in the weeks after arrival: from cleaning and furnishing the house after 6 months' absence to conducting maintenance and repair work in the village, coordinating the shepherds' movements through fallow village fields, and organizing further agricultural activities.[28] And yet, the residents of Son village, with whom I migrated, generally find that the joys of arrival by far outweigh the heavy workload. Many of them haven't seen each other during the winter, and so make use of the evenings to meet and exchange news as well as to collectively

[25] Most migrating people take a jeep up to Sobla, while some village members, often low-castes, bring the animals from the winter settlement to the road-end. The yaks and yak crossbreeds are anyway grazed near Sobla during the winter. Residents of the village Son who do the complete migration on foot usually require 8 days. Starting in the winter settlement Nigalpani the route looks as follows: Dabat, Thawagat, Chirkila, Sobla, Bungling, Sela, Nagling, Son.

[26] Shepherds experience the route between the winter pastures and Dharchula, which takes them between 3 and 6 weeks, as rather exhausting. Major parts are travelled on public roads during the night in order to avoid traffic chaos.

[27] Male cattle and yak-cattle crossbreeds are used as pack animals. Some households also keep poultry, which is carried in baskets during migration.

[28] Agricultural work on the fields starts with the transportation of animal dung to fallow fields that have not been frequented by the shepherds but are meant for cultivation. Usually these are the ones that are directly clustered around a settlement. If available, the manure is transported in jute bags with horses.

commemorate the souls of deceased family members. This latter practice, which is then successively celebrated in each and every household of the village, is known as *syimi thomo*.[29]

The top-right or top-left corner of the kitchen, normally located at the rear side of a house, is the place of the dead souls or ancestor spirits, called *syimi*. It is clearly recognizable by a large bundle of white ribbons (*daja*) fixed to a roof beam. For *syimi thomo* all invited guests gather in the kitchen and wait until the officiating ritual specialist, often in the company of the attendant eldest male member of the family, to initiate the ceremony. They flame some dried branches of *syir sying* (*Juniper indica*), which is widely used as incense in acts of ceremony. While whispering a mantra to invoke the deities' blessings, each of them ties a new *daja* to the already existing bundle. Jointly with all the other participants the two men then toss some grains of uncooked rice and loudly repeat the main part of the mantra saying '*he pamsare rthonya*' ('let us throw the offering'). Following this announcement, the ritual specialist offers some liquor from each bottle that has been provided by the host and his guests. By dipping his forefingers into the spirit and sprinkling a few drops on the ground – something that most villagers, Hindu and Buddhist, do throughout Kumaon – he gives a share to the deities. In a similar way little pieces of the main food items prepared for the evening are offered, before a small plate with food and drink is set apart for deceased family members on a small tray next to the bundle of *daja*.[30] Hereafter the members of the host family distribute liquor, *bakchi* (fried bread), and meat, or, if no animal was sacrificed beforehand, some sweets to their guests.[31] As a blessing all participants further receive a blade of grass and a flower behind their ear as well as a *tilak* (mark on their forehead). The rest of the evening is spent in a relaxed atmosphere: eating, singing, dancing and discussing on the work program for the next day.[32]

When the time comes for the ploughing of fields and the sowing of crops, preliminary preparations are made for a collective village ritual that symbolically

[29] *Syimi thomo* is usually translated as *pitru puja*, or ancestor worship, in Hindi. Through this form of worship people honor the dead and keep them from becoming ghosts. In a very general way it also allows people to obtain their blessings. However, they typically avoid any direct contact with the dead, because this is considered a dangerous and ghostly affair.

[30] The householder usually takes the food that was set apart for the ancestors during the ceremony outside on the next morning. Crows are attracted to make them transmit the food to the departed souls. It is a common Hindu religious practice to offer food to the crows when paying homage to the ancestors, as it is done, for instance, during the so-called *shraddha* ceremonies.

[31] Alcohol is traditionally distributed in a special bowl, called *kuru*, which is passed around several times among the participants. While elderly married women are principally allowed to consume alcohol during such ceremonies, they drink – if at all – much less than the men. Low-caste villagers usually receive their share in a separate bowl.

[32] The favorite dishes on these occasions include potatoes and reddish. Most households store a larger quantity of these field crops from the previous year in approx. 2 m deep wholes in the ground. These wholes are tightly closed with stones and birch bark. When the crops are taken out in the spring, they are kept outside during the day to activate sprouting. The young leaves of reddish are also cooked as a vegetable.

initiates these strenuous tasks. The ceremony is part of a threefold sequence that unfolds until the time of harvest in October. The connectedness of these rituals becomes clearly visible through the establishing, renewing and removing of long poles made from the trunks of conifers (e.g. *Abies spectabilis*) or birches (*Betula utilis* var. *jacquemontii*) that have been de-limbed and stripped off their bark. These sacred prayer-flag poles are locally known as *alam*. While they grace every house as well as all major places of worship, the most noticeable ones are always located at the main village temple.

These huge poles are first installed in May during the worship of the god *Gabla Sai*, who is regarded as the divine ruler and protector of the Darma valley. In June they are exchanged – at regular intervals of presently 4–6 years – as part of a ceremony for *Syang Sai*, a community god worshipped in all three valleys of the *Rang*. After a third spectacle at the time of harvest in mid-September, when a ritual for *Su-Ringding*, the god of the fields and its crops, is performed, the *alams* are finally removed and stored on the temple's roof for the coming year. This threefold ritual sequence markedly illustrates, as already Mauss (1979) observed in another context, that seasonal variations are not external but rather inherent to the fluctuations of social life including the power relations that come along with them.[33]

4.2.1 Establishing Fields of Involved Activity

Approximately a week before the ceremony for *Gabla Sai* takes place, a low caste resident of Son village restarts the flourmill. Every household brings a small sack (about 10 kg) of buckwheat from the previous year's harvest, which is grinded into the flour from which to prepare the special food items consumed during the ceremony. These are two varieties of a conical shaped raw pastry called *syildu* and *dulang*.[34] Most households also start the process of distilling liquor and fermenting beer from ingredients they have brought from Dharchula, especially wheat, rice, jaggery and *balma*.[35] In the process of these activities the villagers increasingly long for the day of the ceremony, especially if good weather indicates that the time for tilling the fields is due.[36]

---

[33] For a discussion on Mauss's approach to seasonality see Harris (1998: 66–67).

[34] While the former are also produced for daily consumption, especially after harvesting in autumn, the latter are solely utilized in ceremonial acts, somewhat similar to the Tibetan *torma*. *Syildu* is typically consumed with a grinded paste made from dried red chilies, garlic, salt and fresh herbs; *dulang* is eaten without any additions. Both are accompanied by local beer (*marti cyakti*) and liquor (*cyakti*).

[35] *Balma* is a starter culture prepared from juniper berries, roasted wheat flour and herbs.

[36] It is important to note that the timing of these rituals is not fixed. Should the weather suddenly change or should other urgent tasks arise, the day can be flexibly shifted after consulting with all parties.

The day of the ceremony is stipulated only shortly in advance. The ritual starts in the late morning.[37] All male residents meet in front of the village's main temple, bringing some amount of *syildu* and beer that is collected for later consumption. The ritual specialist then climbs onto the temple roof and passes down the two huge *alams* of *Syang Sai* that had been stored there over the winter. Before they are positioned between piles of rocks on the ground, the priest attaches a new white flag at the top end of each of them. Upon completion, people proceed to the main shrine (*sai ban*) of *Gabla Sai*, where they install a slightly smaller *alam* that was newly cut in the early morning. In the meantime the village youth, both male and female, visit all the major shrines of the village, concluding at that of its tutelary deity called *Mujir*. At each shrine they attach a new *daja* and in doing so incorporate these divine constituents of the environment into the current of peoples' everyday activities: henceforth the village deities are understood to remain present in the village and to partake in the daily life and fortunes of its inhabitants.

Thereafter all the men go home to put on a white woolen dress, called *ranga*, before coming together with the female and low-caste village residents to a field that is dedicated to *Gabla Sai*.[38] In Son village the family of the former *padhan* owns the field on which *Gabla's* shrine is marked by a huge rock.[39] Only the high-caste male villagers enter the field and line up side-by-side at its centre. Each of them puts a white piece of cloth filled with seeds of buckwheat (*palti* and *bai*), red-dish (*lapho*) and some coins on the ground. The men stand directly opposite the women of the village who – young and old, high- and low-caste – are sitting on the ground at the margin of the field holding tufts of grass and petals in their hands. At the lateral side stand the low-caste males with different sorts of drums, rattles and cymbals. When all have taken their place, the latter eventually impel the ceremony with the commencement of their music.

The men on the field take one variety of seeds from their cloths, turn around and – with loud rejoicings and embraces – toss them away as if sowing, while the women fan in a circular motion their tufts of grass and flower-leaves in the air.[40] As soon as these two groups of actors once again face each other, the women prompt their male counterparts to carry on. The men follow these calls and, with a great deal of passion, take a coin in each hand and turn around again. As the drumbeats grow

---

[37] Worship of deities almost always starts in the morning, whilst ritual practices concerned with spirits, demons and possession happen in the evening or night.

[38] The *ranga* is worn over pants and includes a white waist-belt and a white turban. Nowadays people rarely wear this dress. For *Gabla's* worship, for instance, many men were only wearing the turban.

[39] The British appointed *padhans* as local officers who were responsible for the revenue administration in their villages. These officers were typically remunerated with land assignments called *padhan-chari* (Mittal 1986: 36–37). Although the office does not exist anymore, the title *padhan* is still widely used as an honorific in everyday conversations.

[40] All over North India and also in the Bhotiya valleys it is exclusively men who plough the fields, while all other tasks can be accomplished by both sexes. One reason for this restriction is that the activity of ploughing is associated with male sexual activity (Sax 1991: 26; cf. Polit 2012: 170–182).

louder, the men, themselves cheering, strike their coins against each other before swinging around to their starting position. The whole procedure is then repeated with another kind of seed, and then another, until, in the final round, the coins too are flicked into the field.

After the exuberant crowd has returned to the temple, the male participants enter it, and once again worship the deities of their village whilst distributing the collected *syildu* and beer amongst themselves. With the exception of the low-caste musicians everybody returns home. Following a short break, the musicians start playing again and slowly proceed on the village footpath to visit each and every household. They halt in front of the main doors, drumming and shouting for those who stay inside. The women rush out first, hiding some ghee in their hands, that is smeared over the heads of the musicians amidst affectionate laughter from all sides.[41] After this reception, the drummers take a seat in the courtyard, where they feast on the plentiful liquor and snacks that is offered to them.

In the meantime, the ritual specialist climbs onto the roof ridge of the respective house, where he affixes a birch branch with white ribbons (*daja*) in *Gabla's* name. Following this blessing of the house the householder generously rewards the musicians for their services with rice, flour, oil, ghee, lentils, and bottles of liquor, as well as smaller amounts of cash. Once they have passed every house in the village, they redistribute these items amongst themselves in equal shares. That, however, is a long way off, especially since the intake of alcohol stimulates people to extend their visits with abundant interludes of song and dance right into the late afternoon. By then some young village men will already have prepared the gathering place besides *Gabla's* shrine for the closing ceremony in the evening: a joint meal in which everybody, including the contract shepherds, participate.

At the onset of darkness, each household brings some food and a conical shaped raw pastry called *dulang*.[42] The ceremony is initiated in a similar way as the performance of *syimi thomo*, however, with one exception: it concludes by breaking each and every *dulang* in half.[43] Pieces of the raw pastry are generously distributed, but people only eat part of this rather 'heavy' food and soon start throwing and smearing the dough into each other's faces. The playful mood and enthusiastic atmosphere stops only when there is hardly anyone left who is not tainted in a yellowish color. After having enjoyed the subsequently served main dish, people request the village elders to perform part of what is called *myilu tumo* ('throwing away'). This verbal performance, which is sometimes also done during the ceremony of *Syang Sai*, marks the core of the harvesting ritual and is therefore presented at full length later.

Here, I want to reflect on the ritual actions that are carried out during *Gabla Sai's* worship. In the first part of the ritual the high-caste male Bhotiya residents of the village, as the predominant landholders, take center stage. In former times these

[41] Ghee is clarified butter and widely used all over South Asia for daily cooking and ritual.

[42] This is one of the very rare ritual occasions when no meat is consumed.

[43] This happens in a ritually prescribed way. The upper half of the *dulang* is bended forward – in the direction of the shrine – before it is being broken in the opposite direction.

men would have left the settlement soon afterward in order to bring up further goods from the transit camps in the lower parts of the valley before crossing the high passes into Tibet.[44] A trader's wife would then have independently taken over the responsibility for handling the domestic affairs. Low-caste families were, and to some extent still are, closely associated to a particular Bhotiya household, for whom they conduct domestic chores and agricultural work in return for food and shelter. And it is these people who put themselves in the limelight during the second part of the ritual, when they are rewarded for their work in cash and kind.

During migration everyone – regardless of their caste, ethnic affiliation or gender – eats, sleeps and works together. This is slightly different in the village context, where low-caste people sit, for instance, not around the hearth but on the outer wall near the entrance of the kitchen when visiting a Bhotiya family. In other words, while social inequalities temporarily recede into the background during seasonal migration, they progressively come to the fore when people resume their daily routines in the summer settlement. The ceremony for *Gabla Sai* describes an effective strategy through which such status distinctions and divisions of labor become eventually condensed into meaning and also materially manifest in the landscape with the re-installed *alams*.

The main performance takes place on a field that is owned by a wealthy and influential family of the village. On this field all Bhotiya men line up to ritually initiate the sowing season. They do this by tossing away the seeds of different crop varieties. After each crop variety the performers take coins, which they strike against each other to assure a rich harvest. All women of the village – irrespective of caste – support these activities by circulating tufts of grass and flower-leaves in the air. In doing so they set up specific conditions that should support the seeds' processes of growth. Throughout the summer months they are held responsible to maintain these conditions, for instance by pulling weed and taking a variety of other measures that strengthen the health of the crops and prevent fungal diseases and pests. The male low-caste villagers, who stand at the lateral side of the field, ensure the efficiency of this ritual enactment of sowing with their powerful musical instruments. In the weeks following the ritual, however, it is often them who actually accomplish the physically most demanding tasks *for* the Bhotiyas, such as ploughing.

However, the field of relationships that is established during the ritual is never fixed for good and far from being static. When moving to the high altitude grasslands for pasturing animals, collecting medicinal plants or hunting wild animals, the differences between high- and low-caste persons fade into the background again. And also within the spheres of daily village life they remain, up to a certain degree, flexible and open to re-interpretation. Many migrating low-caste families have become less dependent on their association with Bhotiya households, especially due to the emancipatory income generated through the collection of *kida*. In order to keep their allegiance and make use of their working power they have been granted

[44] The passes were opened approximately by July and many traders undertook two tours during the summer. Today it is only the shepherds who make a move to the high altitude pastures after the ceremony for *Gabla Sai*.

usufructs on agricultural holdings in good locations otherwise not cultivated anymore by the (non-migrating) Bhotiya landowners.[45] This hints toward another aspect of the performance, namely that it also allows people to uphold a sense of unity, trust and care as inhabitants of the same village. Following Ingold (2000: 47), I therefore argue that the Darmanis' practices of agricultural work, including rituals, are grounded "in a direct, mutually attentive awareness of self and other in shared contexts of experience". This is without denying the fact that the established fields of involved activity might become exploitative and fraught with tension. What is crucial to understand is, first, that the degree according to which people play a part in specific arrangements, especially when cooperation is required during work-intensive times, is constantly re-negotiated and, second, that rituals play an important role to this effect.

4.2.2 The Alam*-Exchange*

The second ritual event I describe in this chapter happens after sowing has been completed (Bergmann et al. 2011: 118–122). It concerns the exchange of the two *alams* that are positioned in front of *Syang Sai's* temple. 'Syang' means 'great' or 'major' and 'Sai' means 'God'. Accordingly, *Syang Sai* is often depicted as *Mahadev* or more precisely, as the Hindu god *Shivji*. The planning for the major ritual event of exchanging this deity's *alams* starts 1 year in advance under the guidance of committees, which are established particularly for such purposes. Since approximately one decade individual villages have started to revive these ceremonies as major communal events at 4–6 year intervals.[46] All village members are expected to participate, especially those who reside elsewhere in India or abroad. While many resettled families visit Dharchula quite frequently in order to attend life-cycle ceremonies, such as marriages, this is one of the very rare occasions to gather in one's ancestral village. The number of attendees, even in small villages with less than ten migrating households, easily exceeds 250 people.

I first documented part of the ritual festivities when the residents of the uppermost Darma village, Sipu, celebrated it in June 2008. The non-migrating families with whom I travelled from Dharchula attached posters to tea stalls and shops all

[45] In many villages low-caste families have also been allowed to occupy houses of non-migrating Bhotiyas in return for maintenance and repair work. The houses of low-caste people are usually rather small and located at the village outskirts. Also the agricultural land that is owned by them is most often far from the village and more difficult to cultivate.

[46] As described in the previous chapter, the whole idea of organizing such major village events goes back to the initiatives of young community members during the 1970s and 1980s. As university students and young professionals they regularly used the summer vacations to visit their parents and relatives in the upper settlements. These visits were conducted in small groups that were identified – like the regular migratory groups – as *kuncha*. Now that most of these people work permanently in urban settings, they wish to offer their children an opportunity to make similar experiences and develop feelings of attachment to their ancestral villages.

along the route. These were titled 'Homage to *Shiva*: Let's Go to *Shiva's* Abode'. It had a red *trishul* (trident) printed on it, and the villagers themselves – with the support of hired porters – carried a massive trident as well as a *damru* (two-headed drum) all the way to the high altitude settlement as an offering to the god.[47] The text on the poster was written in the Hindi language and gave information on the regional popularity of *Shivji*. It portrayed a local legend according to which Lord *Shiva* is said to have once traversed the Darma valley on his way to Mount Kailash. When the local gods and goddesses got sight (*darshan*) of him, they instantly prayed that he would choose the valley as his spiritual home (*tapasya bhumi*). As it was further written:

> But Lord *Shiva* ignored them and kept on moving forward. Whilst his attendants created several obstacles along the way, he advanced without much effort. In that manner Lord *Shiva* soon passed the valley's centre on the Danthu-Bidang road. Since *Shiva* was determined to reach Kailash, he kept on walking at a fast pace and all the gods and goddesses gave up their efforts to stop him. Only an ascetic named Hyanjari made one further attempt. By uttering a powerful mantra he gave rise to a heavy storm that blocked all the high passes into Tibet. It is said that due to one of these hindrances *Shivji* suddenly slipped, but in the very moment that he was going to fall Hyanjari gave him a supporting hand. The god was very grateful for the unexpected help and fulfilled the rescuer's request to visit his spiritual hermitage. Thereafter that place was named 'Shivu' and the village is nowadays called Sipu.

While the legend claims a long-term historical continuity, the significance of *Syang Sai* has changed over the last decades.[48] Nearly all of the villages in Darma recently built a temple for *Syang Sai*. These facilities, which had formerly been absent, closely resemble those in other parts of the Kumaon Himalaya. Inside these temples one typically finds a *linga*-shaped stone, a widespread symbol of Lord *Shiva*, besides a birch twig with white ribbons or *dajas*. On its interior walls it is recorded how much money individual persons have donated in specific years for building and maintaining the temple as well as for realizing the communal ceremonies related to *Syang Sai*. Residents say that the building of modern temples helps them to keep traditional cultural practices attractive and understandable, especially to those who lead an urban middle-class life.[49]

During the week of *Syang Sai's* ceremony the village experiences a sudden eruption of life to an extent where the elderly inhabitants say that they are taken back to the past: every single house is occupied; children of all ages run around in the village and through the nearby forests; wherever you go there are people with whom to exchange news, have discussions and awaken memories; and the evenings are

---

[47] Both items are typically associated with Lord *Shiva*. See Sax (1991) for the significance of this deity among the Pahari.

[48] Sherring (1905: 116–117) mentions that whereas *Gabla Sai* was universally worshipped in all three valleys of the Rang, the Chaudansi alone would put their faith in *Syang Sai*.

[49] The traditional shrines of deities are all in the open, besides rocks or under trees. These places are marked with *dajas* and, perhaps, a bell. While urban-dwelling Bhotiyas still visit and also cherish these places, they yet consider it adequate to maintain a 'proper' temple to celebrate the major religious village ceremonies.

enjoyed in convivial gatherings with relatives, neighbors and friends from other villages. In Son village, where I participated in the event in June 2009, the ceremony itself lasted 3 days. On the first day everyone gathers together inside the temple during the early evening hours, where people worship the god and sing songs in his praise till far into the night. The singers' repertoire is large and comprises devotional songs (*bhajans*) in Kumaoni, Nepali and the Darma language. In recent years talented community members began composing new lyrics in their 'mother' tongue, which are introduced on such occasions to the joy of all. An out-migrated person from Son village penned, for instance, the following song dedicated to *Syang Sai*:

> At the time of coming to your door,
> we all felt tired and hungry.
> At the time of your worship,
> we all got ready and came.
>
> We came to Darma in your name.
> We all reached this beautiful land.
> All friends have met through your kindness.
> The people of the village have found each other.
>
> We ate potatoes and radish.
> as well as *palti* and *bai*.
> We enjoyed the greens of the upper jungles
> as well as sprouted buckwheat.
>
> We have worshiped our ancestors.
> *Mu sai* has been honored in the upper house,
> and *Nyundan Sai* in the lower stable.[50]
> We have given a *daja* to each god and goddess.

In such songs local food items and consumption habits are one of the most widely used metaphors to convey the 'flavor' of a perceived cultural heritage. This highlights how the relations in which people engage during these days "are not only moral and spiritual but also and at the same time corporeal and substantial" (Sax 1991: 72). It is particularly by eating and sharing locally grown food items that people re-activate a sense of belonging to their ancestral villages. While such a substantialized dialogue with one's material environment is a widespread phenomenon in South Asia, it has special corporeal entailments for the Darmani. This became clear, as I will now describe, during the main part of the ceremony on the following day.

Early in the morning the ritual specialist brings a young fir to the temple, which he cut from a nearby forest. After arriving he ties a white piece of cloth to the tree and tosses some grains of rice on it. Through this common practice, called *daja chimo*, the fir is transformed into a ritual object classified as *dhuksying*.[51] Shortly afterwards a sheep is sacrificed within the temple, and in the presence of all male

---

[50] *Mu Sai* is typically translated as *Kul Devta* and *Nyungdan Sai* as *Kul Devi* in Hindi, i.e. the family god and goddess.

[51] Some of my informants have related the *alam*-exchange to a marriage custom in the neighboring Byans valley, where the party of the bridegroom carries along a similar tree, called *darjyo*, when setting out to collect the bride.

village members, whereupon people of both sexes gather in front of *Syang Sai's alams*. Eventually they begin to dance in a long row toward the lower entrance of Son. This style of dancing (*choliya*) is common on many occasions that require an orchestrated movement away from or toward the village centre, especially for the reception and farewell of important village guests or relatives. Led by the ritual specialist carrying his small fir, or *dhuksying*, and accompanied by a small group of low-caste musicians, the alternating male and female dancers only halt after they have reached the main footpath of the valley below Son proper. From there the male participants advance alone toward the next lower village, Baling, where they again line up to proudly dance past its residents.

The final destination is a nearby forest patch with particularly large firs. Although the area forms part of Baling, some of the upper villages hold usufructs – obtained through kinship relations – to extract the forest resources required for these ceremonies. Before the villagers proceed into the forest to pick out suitable trees, the ritual specialist is required to exchange his small fir (*dhuksying*) on the fringe. For that purpose he cuts a tree of similar size, fixes a small piece of white cloth on it and puts the old one aside as to remain clearly visible for any passer-by on the wayside. Within a short time, people then log two towering trees, which they thoroughly de-limb and strip off their bark except for the fruit-bearing crowns. Since only two persons are permitted to shoulder them at the same time, the homeward journey is by far the most demanding part of the whole ceremony. While the movements before were displayed as an elegant dance, they then turn into a parade of physical bodily strength. Only by relieving each other every few minutes are the participants of this procession able to safely navigate their heavy loads in this difficult terrain (Photo 4.3).

When the party is about to pass Baling, the residents of that village welcome them warmly and serve a quick meal and liquor to their guests, who individually reciprocate this hospitality with smaller amounts of money. This mode of exchange, called *chyaitan*, happens at the lower entrance of Son village again. There the men's arrival is celebrated even more joyously than before, until male and female villagers dance together again, following the *alams*, to the main temple. At the temple the women group themselves in a half circle. While their male counterparts first remove the old (*alam kenya*) and then install the new *alams* (*alam chumo*), they ceaselessly throw rice on the ground and fan in circular motion white pieces of cloth with their hands. Both gestures encourage the god to accept and occupy his new seat. The exchange (*alam sammo*) is completed after the ritual specialist has put his small fir (*dukshing*) in *Syang Sai's* shrine and purified the temple's roof with spring water and cow urine. The overall ceremony is concluded on the following day, when the village community sponsors a splendid meal at the temple to which relatives and friends from all over the valley are invited.

The performance describes a ritualized movement that is clustered around biotic resources of scarce distribution in higher altitudes: massive tree trunks. Radcliffe-Brown (1922) once asserted that the prominence of such 'natural' objects in rituals reflect their functional-technical importance for a society, including their particular value as a resource. According to Radcliffe-Brown such values bind the individual

Photo 4.3 Carrying the *alam* (Photograph © Christoph Bergmann, June 21, 2009)

members of a society and rituals facilitate the communication and conservation of corresponding interests in a standardized symbolic form. In the present case this could be formulated, for example, as 'firs are widely required construction materials but a scarce resource, and the ritual shows their value'. One might indeed be tempted to give a functionalist explanation, especially when taking into account that the pressure on forest resources was even stronger in earlier times: in some villages the number of migrating people was more than twice as high and the number of livestock – especially of sheep and goats – even up to ten times higher than it is now (Garbyal et al. 2005). Moreover, a respectable number of Tibetan nomads seasonally entered the valley with their animals until before the closure of the Indo-Tibetan border.

Some of these nomads, the so-called Darma-Khampas, once occupied a seasonal settlement that was beneficially located along the former trade route in the uppermost part of the Darma valley, above the timberline. Even though the British did little to regulate their extraction of forest resources, the interference of 'Tibetan' user groups was in general seen as creating both political and fiscal problems (Lamb 1989: 366–372). According to Sherring (1974 [1916]: 86–87) the presence of prayer-flag poles in Taklakot, a major trade mart of western Tibet that was regularly frequented by the eastern Bhotiyas, marked a visible cultural similarity between the Tibetan nomads and the Bhotiyas. Although the details are difficult to reconstruct, the negotiation of access rights to forests in the Darma valley was previously closely linked to the *alam*-system. This system provided a ritualized framework through which the prestigious flow of biotic resources across the altering potentials of an altitudinal gradient became regulated. As a fine-tuned mechanism it allowed different actors to strategically manipulate a resource network that was also lucrative in terms of marriage and trade alliances.

Until today the usufruct rights for the forested areas in which the *alams* are extracted are conceded through kinship relations: if a lineage from a village in the lower parts of the valley gets one of its daughters married to a man from a lineage in the upper parts of the valley, where forest resources become increasingly scarce, the members of the latter lineage can take this as an opportunity to negotiate such use rights. The resulting flow of resources, then, follows a widespread pattern, namely from wife-givers to wife-takers. The Darma-Khampas, who did usually not intermarry with the Bhotiyas, could in principal also participate in these transactions, for instance by entering into commercial partnerships with Bhotiya traders.[52] Without presuming the existence of any homeostatic equilibrium, the general functioning of the *alam*-system seems thus to confirm Rappaport's (1967, 1968) idea that rituals are an important and a complex means to regulate the use of environmental resources.[53]

But this idea does not account for the ongoing significance and attractiveness of the *alam*-exchange as *the* major communal village event. People themselves lay emphasis on the distinct experience in performing the ritual, particularly of carrying home the heavy trees over a quite respectable distance. This kinaesthetic dimension evolves as a kind of male 'muscular consciousness' that becomes manifest with the installed *alams*, propagating masculine strength and potency among a (re-)united village community (Ingold 2000: 203). It even allows the participating urban dwellers to experientially associate themselves with the hardships of a mobile agro-pastoral life that is perceived as a crucial ingredient for cultural heritageization. However, this positive attitude goes hand in hand with rather skeptical assessments of local residents who worry that the excessive felling of trees for ritual purposes may potentially cause harm to the environment (Bergmann et al. 2011: 121–122). Such assessments are understandable if one considers that the *alams* are not only installed at temples and shrines but also at each and every house that is at least temporarily inhabited during the summer months. In the year of a major village ceremony, people do thus require a considerable number of trees.[54]

---

[52] In August 2009 some Darma-Khampas visited their former settlement Khimling (also called Bidang) for performing a ritual. They as well carried along an *alam* from Baling village. They, however, had paid for the tree.

[53] Rappaport argues that rituals are crucial for the maintenance of a society's balance with natural resources. According to this system-centered perspective, which has received renewed attention recently and is once again prominent in environmental anthropology; the distinguishing features of rituals are complex homeostatic culture-nature feedback mechanisms (cf. Biersack 1999; Messer and Lambek 2001). Through these mechanisms the distribution and conditions of biotic resources as well as the relationships between social groups are thought being regulated in empirically measurable ways.

[54] It should, however, be noted in this context that the exchanged *alams* are further used, for instance as firewood.

Agrawal (2005a: 166) adopts the term 'environmentality' to foreground the "technologies of self and power [that] are involved in the creation of new subjects concerned about the environment", such as the allocation of financial support and the spread of 'expert' knowledge. However, his approach leaves unspecified the forms of regulation that are practiced without state recognition and without compliance to the rationalities of modern governmentality. Moreover, as Singh (2013) argues, this framework neglects the role of affect, emotion and embodied practice in shaping the subjectivities of people who care for their environments. So far the forest councils as well as other legal institutions in the Darma valley are not detached from but rather are entangled with various informal institutions such as the *alam*-system. I contend that the *alam*-system has the potential to support and complement legally established procedures of resource use and sharing, because it enables people to better integrate detached and disembodied management rationalities with kinaesthetic and embodied practices. Taken as an informal mode of communal forestry the *alam*-system amounts to a mode of environmental engagement through which the conditions of forest resources are not so much measured as they are felt. Powerful deities, such as *Syang Sai*, mediate this experience. While the spread of expert knowledge on environmental sustainability has made some people skeptical about the reasonableness of cutting trees for ceremonial purposes, I have met nobody who aspires to really abolish such activities. This is because the *alam*-system describes not so much an activity through which people intend to act upon and transform an external and quantifiable 'natural' landscape, rather than an opportunity to actively participate "in the currents of a world-in-formation" (Ingold 2011: 129). It is by getting immersed in these currents that people bring themselves and their environments continually into being (Ingold 2000: 87; Taylor 2015: 21).

4.2.3 Celebrating Harvest

In mid-September, when the shepherds gradually return from the alpine pastures to graze their animals on harvested or fallow fields as well as on grasslands in the valley bottom, *Su-Ringding*, the god of the field crops, is worshipped (Bergmann et al. 2012). During this time a variety of conflicting tasks remain pending, such as crop processing, wool shearing, and preparations for downward migration. Flock-owners count their animals in order to determine payments for employed shepherds, and the low-caste residents are reimbursed in cash or in kind for their labor contribution.[55] While the diverging demands and loyalties of the various actors often give rise to conflict, the ritual foregrounds how people harnesses the seasonal variability of their dwelt-in surroundings to channel potential tensions and conflicts away from the spheres of settled life and seasonal migration.

---

[55] For wool-shearing a person receives approx. INR 10–15 per animal. Crop processing is usually paid in kind.

The ceremony starts early in the morning, when male villagers offer a small quantity of grain from the new harvest at the deity's shrine. In the meantime, the women of every household prepare some food from local grains and pulses. After a joint meal with one's own family, people begin to stroll around in the village exchanging further food and drinks, a joyful activity that goes on for some hours. Eventually, everyone comes together at a place near the deity's shrine, where a black goat is killed with a single stroke through its neck. Diluted with water, the animal's blood is directed along the sloped footpath towards the lower village boundary. This way of killing is exceptional, as people normally rip out the animal's heart through a little cut in the abdomen without any blood leaving the animal's body. Whenever an animal is beheaded during a ceremony, it is not considered to be a sacrifice for a deity but it constitutes an act that should satisfy and ward-off evil spirits (*syino*). Having received their share, these beings are expected to stay away from the village where they are prone to spread unrest, conflict and disease. This clearly is the case when people celebrate the new harvest.

Having cut up the meat, people start to collect some flour, edibles as well as liquor from every household that is put in front of the shrine. Furthermore, every household holds a small mat made of straw (called *kayo*) ready on which some of these materials will be placed at the end of the ceremony. For the time being, however, people eagerly await the arrival of an elderly male villager who conducts a further ceremony at nightfall. This man sits down in front of a wooden bowl or some other vessel that is filled with parts of the offerings, a small piece of the goatskin as well as some meat. When he starts to perform what is called *myilu tumo* ('throwing away'), low-caste drummers accompany him. The performance is constantly interrupted by loud comments from the audience's male section. All along they litter flour and distribute liquor amongst each other while the elderly person – holding a wooden stick with a burnt tip, with which he fiercely mixes the ingredients of the vessel – recites the following story[56]:

> The gods and goddesses of my village are quite ordinary.
> The villagers, also, are somehow not different.
> Looking at their flocks, they seem rather poor.
> Also their crops do not grow very well.
> Among their sons not a single one is son-like.
> Among their daughters not a single one is daughter-like.
> None of their belongings seems to be special.
> But in the village in front, everything is better.
> Good houses, fertile fields, handsome sons, easily graspable daughters, huge flocks, and many fine goods are abundant.
>
> If you must get jealous, have envy or malice, then save it for those people! And if you have to go, then leave in that direction. But hey, those from the lower villages do the same, and they also send their people to the next village. And as this continues, the last of our villages will finally recite these verses for the whole of our beloved Darma region.

[56] I documented different versions of this ritual in September–October 2008 and 2009, namely in village Bon, Son, Dugtu and Danthu in the upper Darma valley. The description is based on the performance observed in village Bon in October 2009. Parts of the ritual chant are accessible in the Darma language in Dhakriyal (2004a: 253–238).

Bad spirits and demons keep looking upon the gods and goddesses, and upon us who dwell in peace, prosperity, happiness, love, and with all virtues in the fourteen villages nestled in this Himalayan valley, this quiet heaven, this golden land. With jealousy, they always try to disturb and spread unrest. This envy, jealousy and malice, all bad words and conspiracies, all dangerous diseases, bad ideas and thoughts, tiredness, the origins of all kinds of violent outbreaks, power fights and ill effects we now send back from here.

Today all these evil impacts will be shed into the flowing river to reach the far-away and deep waters. But if this evil packet indeed goes down with the rivers into the deep waters, it will clash with those huge elephant-like beings residing there.

These giant creatures that, before discovering this evil packet, will have been quite amused, will of course ask, "What in the world is this?" These powerful beings will be affected by those evil influences and in a sudden burst of anger they will throw this damned packet high into the air. But guess what will happen to it? It will be carried along by the wind and again reach our beloved Himalayan glaciers.

All of a sudden this evil packet will fall out of the sky and land on one of our high pastures where our yaks are grazing. Think about how this packet will look like in such a place where one cannot find even a single blowfly. Having seen this evil packet, also the yaks will get furious and produce loud noises with their noses and leap about on their four feet. But finally these fearless beasts will come close and pick this packet up with their horns. With all their strength they will again throw it back into the deep rivers.

In the closing part of the performance, the evil package is verbally directed through every one of the summer settlements and several other stations lying on the path towards the winter settlements. Its final destination is a place called Bakrihath, which is located south of the last winter settlement of Jauljibi near the confluence of the Kali and Gori rivers – the 'deep waters'. Jauljibi hosts an international trade mart for livestock and commodities, which takes place every year in January. In former times local aristocrats, presumably the 'elephant-like beings' mentioned in the ritual chant, lived there and were supported by their most loyal subjects. Worth mentioning, for example, are the Rajwars from Askot, local aristocrats who held proprietary rights in the village lands and harshly restricted the Darmanis' access to cultivatable fields in these lower parts. Darma shepherds still complain that their position beyond Jauljibi becomes increasingly insecure due to more unreliable social alliances, thefts of livestock, corrupt forest guards and the like. In a heightened atmosphere, the village elder proceeds as follows:

Give voice, make abuse
Throw it away, I say!
Give it away, I say!
The wood taken from the glowing fire is not going?
Far from it, certainly the wood taken from the glowing fire is going!
It is going from Sipu – the village of *Syang Sai* – hey!

[After every verse, ending with the name of the village and the tutelary deity of its residents, the audience begins to dance and shout]

We have done the worship for Su today.
Throw the evil packet away, I said!
After six months, after 160 days,
you neither have to say yes to someone,
nor do you have to do something.
If someone still does, then a six-inch nail goes right into his head

The wood taken from the glowing fire is going from Tidang – the village of *Chung Sai* – hey!

We have done the worship for Su today
After six months, after 160 days
you neither have to say yes to someone,
nor do you have to do something.
Admittedly our houses look like corn mills,
but our fields are rather tiny!
And our girls are just like a *tokar*.
And our boys are just like a *takuva*.[57]
The wood from the glowing fire is going from Darkar – the village of *Riya Sai* – hey!

Through variations of these lines and by naming every village along with its tutelary deity the route toward Bakrihath is completed. One male member of each household then takes the small straw mat (*kayo*) on which some of the collected food and drinking items are put. These offerings are either placed besides the shrine or on a nearby footpath from where crows are expected to pick them up and transmit sustenance to the ancestor spirits (*syimi*). In the meanwhile another man – normally a low-caste person whose daily task is to look after the villagers' cattle – approaches the scene and takes over the vessel and the wooden stick from the main performer. Equipped with branches of a prickly shrub as well as cow dung, the village youth drive him nine times around the shrine and then finally toward the Darma River or one of its tributaries. They are expected to 'kill' the low-caste person, who however always manages to escape and to throw away his baggage into the flowing water. The shouting crowd on his heels then deviates from the intent to 'kill' and instead gets rid of its 'weaponry' on the riverbank. When everyone has returned to the village, the person who was in charge of the evil packet is applauded for his bravery and receives the two front forelegs of the killed goat.[58] Afterwards the whole village joyfully dances and sings late into the night.

In view of the Darmanis' pastoral practices the performance emphasizes two interrelated points. On the one hand it foregrounds that the integration of different labor demands as well as the scheduling of divergent agro-pastoral tasks before downward migration is prone to all kinds of social conflict. In the ritual people encounter potentially uncontrollable and life-threatening forces, which they channel from the summer settlement towards the outside of their core migratory route, i.e. beyond the last winter settlement. While village purification rituals are extremely common all over the world, a distinctive feature of the one conducted by the Darma villagers is that the evil influences are channeled all along their ordinary migratory route. The participants acknowledge that they will never get rid of potential tensions and conflicts for good. Like their own migrations between summer pastures in the upper mountains and winter pastures in the Himalayan foothills, the movement of the 'evil packet' is subject to seasonal fluctuations and thermal conditions. Once

[57]The *tokar* is the lower end and the *takuva* the upper stick of a spindle device known as *takli*. The metaphor has clear sexual connotations, referring to male and female genitals.

[58]The animal is either sponsored by some wealthy village member or through the village fund. The remaining parts of the meat are auctioned after the ceremony.

these conditions are suitable again this packet is expected to come back with the winds and turn up again on the high altitude summer pastures.

On the other hand the ritual draws crucial distinctions between different kinds of livestock. Flocks of sheep and goats are associated with wealth, cattle are linked to the conflicts of village life, and yaks are associated with purity, strength and the reproduction of social order.[59] I was often told that in earlier times people recognized the wealth of a family by the size of their flock. While things like education and employment are nowadays often considered to compensate for a lack of these animals, flock ownership and herd size is still considered as the most authentic expression for estimating class equivalence.[60] Things look different with regard to cattle. While the number of cattle, especially of cows, has steadily increased among the Darma villagers over the last decades, cattle herding is not at all hold in a high esteem and usually done by a low-caste village member. In the ritual it is also him who has to take away the evil packet to throw it into the river at the 'risk' of being 'killed'. Yaks, on the contrary, are conceived as willful and wayward agents themselves. Within the Darmanis' pastoral system they are mainly kept as transport animals during seasonal migration and for breeding purposes. For most of the time these animals remain on high pastures, and if the terrain is suitable they are left on these high grounds even without any herder. Also in the ritual they play their own part in rounding up the annual cycle by throwing the evil packet back into the river once it has reached their alpine pastures with the winds. In the mortuary rites of the Darmani, to which I turn now, these animals play an important and agentive part as well, but under different auspices and with a rather different outcome.

To sum up: so far, the ethnography has shown that rituals constitute an integral part of people's agro-pastoral work activities in the Darma valley; that mobility-related rituals allow community members to identify with and physically experience the hardship of a mobile pastoral life; and that seasonal environmental variations are ceremonially harnessed to coordinate shifting demands of labor, to entrench positions of power and to temporarily channel emerging tensions and conflicts towards the outside of the migratory cycle. For the remainder of this chapter I focus upon another set of ritualized activities, one that interrupts the characteristic pattern of day-to-day life and is performed after the cremation of a deceased person has taken place.[61] While these activities also involve forms of movement, the environment is incorporated into a bodily experience not through muscular entailments but through narrative techniques and a temporal inversion of important social relations, namely those between wife-givers and wife-takers. In the following I show how these narrative techniques, as modes of dwelling, allow people to deal with the phenomenon of death along routes of travel.

---

[59] This differs from a nomenclature reported in Stellrecht (1992), where sheep and goats are associated with high altitude pastures and ideas about purity. In the Dolpo region of western Nepal, however, yaks are the culturally most favored livestock, which also figures prominently in rituals (Bauer 2004: 25–38).

[60] See Kapila (2008: 125) for a similar observation amongst the Gaddi of Himachal Pradesh.

[61] For a description of the cremation rite see Leder (2003: 109–112).

4.3 Dealing with Death Along Routes of Travel

Already Nick Allen (1974: 7) observed that many of the Tibeto-Burman speaking communities in the Himalaya, including the eastern Bhotiyas, "have focused their maximum ritual emphasis on death" and that their mortuary rituals typically involve a certain kind of journey. According to Allen the distinguishing feature of these journeys is that they do not involve physical but verbal or imaginary movements through the familiar landscape. While this observation has been confirmed and expanded through further research conducted both in different parts of Nepal (Bickel and Gaenszle 1999) and in the northeastern parts of the Indian Himalaya (Blackburn 2005, 2008), a follow-up study in Kumaon has so-far never been pursued. Previously the eastern Bhotiyas referred to these rites as *gvan*. During the reformist movements after India's independence they were renamed as *sarat* throughout the whole region.[62] Nawa (2007) reports that *sarat* is substantially different from the original *gvan*-ceremony amongst the Byansi residing in far-western Nepal. However, as I shall argue, there is yet some form of continuity in Kumaon.

The *gvan*-ceremony had given rise to various controversies in the region since colonial times, especially because it was associated with yak sacrifices. Since such acts are considered equivalent to the killing of a cow, this means the breaking of what is conceived to be a major religious taboo amongst Hindus. British officials and European travelers reported on this practice again and again (Traill 1992 [1832]: 118; Landor 1899: 112–123; Sherring 1905: 111–115), which then also triggered heated debates both among the Bhotiyas and between them and other resident groups.[63] Already under colonial rule the yak was in some cases substituted with sheep or goats, a practice that gained broad acceptance after India's independence.

In the Darma valley *sarat* is a conglomeration of various ritualized activities that proceed over a period of 3 days.[64] While the *gvan*-ceremony was exclusively performed in the upper settlements, *sarat* is also, or even predominantly, performed in the villages around Dharchula. In case a person dies in any place other than the summer settlement but the relatives decide to accomplish the mortuary rites in the summer settlement, they are required to carry a small bone (*cye*) of the deceased person to that location. Since the carriers of the bone are not allowed to use any path that runs directly above a temple or a deity's shrine, many inconvenient detours need to be taken. People consider the newly departed soul (*nu syimi*) too weak to encounter a deity's presence. It would become too attached to people's living environment and simply not be able to continue its way thereafter. Wherever the way becomes difficult as well as when crossing a river the relatives mark the path to the

[62] The term *sarat* presumably derives from the Hindi word *shraddha*, meaning 'sincere faith'. People commonly use that term in the valleys of Darma, Chaudans and Byans as well as in the Byansi villages of Changru and Tinkar in far-western Nepal (Nawa 2007: 266, footnote 8).

[63] While the word for 'yak' in the Darma language is *gal*, it is commonly addressed as *ya* in the funeral rites. The same name is used if a sheep or goat substitutes the yak.

[64] The original *gvan*-ceremony is said to have originally lasted for 9 days.

ancestral village with white cotton threads in order to assure that the newly departed soul finds its way safely.[65]

Once the upper settlement is reached, people store the container in which the bone was carried in a particular place such as a tree, a garden wall, or a small hole in the ground. It is always located somewhat away from the village center and its main temple (Leder 2003: 112). The soul is believed to roam around in the village until it is called by the beating of a drum that announces the starting of the final ceremonies. On the first day of *sarat* all villagers visit the house of the deceased person's family to bring some wood, grains, alcohol as well as an appropriate sum of money. This practice is called *sying gaisyimo* ('wood contribution') (Nawa 2007: 274). The agnates reciprocate these donations with a meal in the evening, after which all attendees receive a handful of toasted grains. Toasted grains, which have lost their ability to germinate, are widely associated with death (Leder 2003: 113). The feast thus implicitly involves a distribution of the dangerous influences that arise due to the occurrence of death amongst the invited guests.

Both in *gvan* and in *sarat* a special role is attributed to the deceased person's 'sisters', or to the 'sisters' of her husband if the deceased person is female. In the present context the term 'sister' (*chame*) refers to the consanguine female kin within the deceased person's patrilineage. Besides grains and other eatables they collectively provide at least one sheep or goat that is sacrificed toward the end of the ritual. It is expected that their husbands, i.e. the deceased person's sons-in-law, brothers-in-law, and uncles-in-law, accompany them. Those men are obliged to take on all kinds of tasks related to the guests' provision with food and drinks or the organization of overnight accommodation. This group of male and female relatives, which bridges the divide between wife-givers and wife-takers, is addressed as *mya-chame* (Sherring 1905: 113).[66] The (eldest) son-in-law, or in case there is none, the (eldest) son of the deceased person's sister, is traditionally considered the master of ceremonies. His main duty in *sarat* is to prepare ritual items as well as to undertake tasks related to the ceremony. In the former *gvan*-ceremony he would have fetched the yak and, thereafter, escorted the animal through the village until it would have eventually accompanied the newly departed soul to the land of the dead (Allen 1974: 12–13).

For that purpose the son-in-law would have taken the yak through the village on two-to-three consecutive days (Leder 2003: 113–114). On the first day the yak, whose sex corresponded to the gender of the deceased person, was taken to the place where the deceased person's bone was stored. There the animal's horns and feet were decorated and a colored line was drawn from its tail to the neck. Some clothes of the deceased were then tied on the back of the yak and a white cloth – called *am lugara* ('way-cloth') – fixed between its horns (Allen 1974: 12–13).

---

[65] This practice is called *am tanmo*, which literally means 'to make a way'.

[66] *Mya* means 'son-in-law' and *chame* means 'girl' or 'sister'. If the deceased person is male, his mother's brothers are distinguished as *pohya*, a term that is not used outside the ritual context. Attendees that neither fall under the category *mya-chame* nor *pohya* are called *puthye* (Allen 1975: 92, footnote 4).

The soul was then invited to take this animal for its journey to the land of the dead. The deceased person's 'sisters' subsequently fed the yak before the male agnates brought it to a place outside the village, where they tossed toasted grains over it to nourish the soul. On the second or third day the animal was shown around to collect the bone of the deceased and bring it to the house of his family, where the journey of the soul to the land of the dead (*syimi tankval*) would have then been performed. There are various interpretations about what happened to the yak after the completion of that journey: it might have been chased away to the upper mountains and left to care for itself; low-caste village members or Tibetan nomads killed the animal and consumed its meat; or it was driven down a steep rock slope and died. In any case, the ritual is obviously very complex, especially with regard to kinship relations and its geographical layout. However, before presenting a local interpretation and my own commentary to it we shall first have a closer look on the manner in which people do nowadays perform *sarat*.

When compared to the *gvan*-ceremony, *sarat* appears in many ways as a simplified version of it. Male relatives collect the bone of the deceased on the morning of the second day of *sarat*. They do this with a special cloth-bag, the *syimi babchai* (soul-boot) sewn by the son-in-law the night before. The soul-boot, containing the bone, is brought to the house of the deceased person's family and put up in a corner of the kitchen. Below it the so-called *yan* is established, which consists of a (wooden or plastic) box with a large, flat stone on top. On the rear side of this shrine a small slot is kept open, into which the abundant offerings of food and drinks are thrown after having been 'consumed' by the departed soul. The stone itself is covered with a beautiful rug on which a photograph of the deceased, some incense and an oil lamp is placed. The forefront is decorated with precious jewelry of the deceased person's sister. In the *gvan*-ceremony the son-in-law would often also have prepared an effigy of the deceased that was set-up besides the *yan* (Leder 2003: 115). This effigy consisted of a wooden cross of two birch tree branches decorated with the clothes and other utensils of the deceased person. Nowadays, however, the photograph is widely taken as an adequate substitute.

Towards the evening of the second day all participants come together inside the house to undertake the most challenging part of the ceremony. The ritual specialist conducting the ceremony is called *amricha*. In general these are men, but in one of the instances I documented, it was a woman from Bungling village. More important than the person's gender is that he or she knows how to guide both the bereaved in mourning and the departed soul on its way to the land of the dead. The name *amricha*, which literally translates as 'path-viewer', is telling in that regard. He is always assisted by a second person, the *banricha* (place-viewer), who takes care that each and every place is frequented in the correct order and that no place is left out on the way. The major skill of the *amricha* is the proper completion of an oral recitation that continues – with short breaks in between – for 12 h or even longer.

Major parts of the oral texts chanted are called *amrhimo*, a term that can be translated as 'route survey'.[67] It is in the recitation of this text that villagers see the continuity with the former *gvan*-ceremony, even though these oral texts have changed over the last decades. According to Nawa (2007: 266–273) these texts are not anymore recited during *sarat* amongst the Byansi in far-western Nepal. Instead a Tibetan monk recites some Buddhist text, which is followed by the singing of Hindu devotional songs (*kirtans*). From the material I collected there is no doubt that *amrhimo*, as it is nowadays sung amongst the Darmani, has been modified with references to mainstream Hindu motifs and practices. Elderly people variously told me that *amrhimo* was different in earlier times. However, I do not want to dwell on these modifications here but leave it open for further analyses. My objective is rather to present part of the recited text and to analyze the manner in which it channels the emotions of and forges the relationships amongst the bereaved.

Whilst educated community members tried to write down an authoritative account of these oral texts (Dhakriyal 2004b), the small category of persons acting as *amricha* do still not follow a single canonical version.[68] The storyline is fixed in that it follows certain prescribed paths of travel, but yet describes an improvisatory process in that each performer creates a specific accentuation. The ritual chants narrate the manner in which the gods and goddesses of the Darma valley have dealt with the phenomenon of death and the disturbances it creates. The whole recitation starts at nightfall. The bereaved sit inside the kitchen, weeping and crying for the deceased. Some low-caste drummers have gathered outside the house with their instruments, awaiting the ceremony to start. This happens when the son-in-law, accompanied by the 'sisters' of the deceased person (or of her husband, if female), comes out with a burning piece of wood taken from the hearth. While the drummers set in with a furious rhythm that is described as 'inversed' (*namphan*), the female relatives light small torches and start dancing in the courtyard according to an order of steps that is equally considered as an inversed pattern. When people then return back inside the house, the *amricha* has already initiated his oral recitation with the following lines:

[67] Over the last decades various discussion have taken place with regard to the funeral ceremonies and in many places villagers experimented with alternative forms. Some, for instance, engaged Brahmin priests to recite parts of the Sanskrit *Garuda Purana*. Whilst such attempts did, however, not gain acceptance, some people still argue that the oral texts of *amrhimo* themselves describe a kind of *purana* or ancient Vedic text. Dhakriyal (2004b: xx), for instance, speaks about the *Kak Purana*.

[68] In order not to evoke the dangerous forces that are released during these performances, people do usually not allow the ceremony to be recorded. The residents of Son and Dugtu made an amiable exception in July 2009. The situation was particularly tragic at that time. A young boy had died in a car accident on his way to Haldwani, where he was invited for a job interview. I am very much indebted to his parents and relatives for consenting to my full participation and recording of his funeral ceremonies. The description of *sarat* provided below follows the proceedings I observed in Dugtu. The Hindi translation of the text was produced with the help of Dabla Singh Phirmal, who acts an *amricha* himself. As some passages proved rather difficult to translate, we partly resorted to Dhakriyal's (2004b) account.

For ages our forefathers have followed the rules that were established at the time of the world's creation.
Nobody has the right to create unrest and to breed bad blood in this golden land.
Nobody is allowed to doubt the fertility that flourishes in this golden land of the Himalaya.
No single man is allowed to keep his head bare.
No single man is allowed to beat his head with the palms of his hands.
No single man is allowed to beat his chest in this golden land.
Nobody is given any chance to drop as many tears as raindrops fall during the monsoon.
No married woman is allowed to remove her earrings.
Our beautiful sisters are not allowed to remove their bangles from their wrists [a sign of widowhood].
No chance is given to our wives, sisters and daughters to leave their hair loose [a sign of menstruation].
None of our sisters has the right to wear her scarf backwards [a sign of mourning].
The son-in-law is not allowed to light a torch and to move around with it in the village.
Nobody has the right to distil liquor for inauspicious purposes in this golden land.
Nobody has the right to set fire and gather around the hearth in these times.
The musicians have no right to play the inversed beat with their drums.
Nobody shall search for the yak on the upper pastures.
Nobody shall escort the yak through the village.
Nobody has the right of feeding the newly departed soul.
Nobody has the right to distribute nine types of toasted grains.
Nobody has the right to blow the copper trumpet.
Nobody has the right to distribute liquor.
Nobody has the right to request the son-in-law to sew the soul-boot.
Nobody has the right to lift the bones of the deceased.
Nobody has the right to dance in an inversed sequence.
Nobody has the right to articulate any inauspicious sentence from his mouth.
Nobody has the right to carry out *sarat* in this golden land.
But today these fortunate rules have to be broken.
Call it good or call it bad, *amrhimo* must be done.

The message is clear: the ordinary codes of conduct are suspended. Inside the house every participant offers some food to the newly departed soul or *nu syimi* before sitting down to listen to the *amricha's* ongoing recitation. Liquor is distributed; plates with cigarettes are passed around; and the atmosphere fills with sadness. Those who find their voice repeat the closing parts of every line chanted by the *amricha*. His assistant, the *banricha*, throws toasted grains in the direction of the *yan* after every completed verse and quite soon the floor becomes coated with these tokens of death, grief and mourning. After each major branch of *amrhimo* the musicians outside the house set in with their inversed drumbeat. Snacks are distributed and some people go out to breath fresh air. But these pauses are brief. After 2-to-3 h of chanting, the *amricha* begins with a narrative that tells the story of the god *Rangchim Sai* and his son, *Mangla Syiri*. Whereas *Mangla Syiri's* mother is never mentioned throughout the whole text, his name gives some information in that regard. *Mangla* is the Darma word for Hindi *man*, meaning 'mind', 'heart', 'soul' or 'character'. In Hindu cosmology, however, *Mangla* is also used as a title for the Hindu goddess *Parvati*, the wife of *Shiva*. *Syiri*, on the other hand, means 'son' in the Darma language, so that *Mangla Syiri* could be translated as '*Parvati's* son'. Such a translation is supported by the fact that the Darmani worship *Rangchim Sai* as an androgynous deity (Sherring 1974 [1916]: 71) whom some people even explicitly associate with *Ardhanarisvara*, the androgynous form of *Shiva* and his consort *Parvati*. Be that as it may, *Rangchim Sai* is half-male and half-female and as such personifies both parents of *Mangla Syiri*.[69]

The story of *Mangla Syiri* is a key passage of the text with which the relatives of the deceased very much identify. In order to understand why, let us follow the winding trails of the narrative itself:

> Once a most beautiful baby named *Mangla Siri* was born in the golden land of Darma. Everyone became deeply affected that such a beautiful baby was born as the son of *Rangchim Sai* from Tidang village. *Mangla Siri* was as lovely as the sun and the moon. Days, months and even years passed in which the boy grew up with an incredibly charming face. But one day the boy came to suffer from a disease. *Rangchim Sai* kept trying different remedies, but nothing helped. Finally, *Rangchim Sai* prepared some white ribbons [*dajc*] and some uncooked rice as well as some food, and then left his home to ask for advice among the gods and goddesses.
>
> First he summoned *Syang Sai* from the uppermost village Sipu. He told him that his son had not taken anything – neither water nor rice – for almost three days and straightforwardly asked about the serious nature of the situation as well as the name of the disease from which *Mangla Siri* was suffering. But *Syang Sai* from Sipu village confessed that he had no idea about this problem and thus recommended *Rangchim Sai* to visit *Mawar Sai*, the god of Marcha village: "He will tell you what has happened to your beloved son". Hence *Rangchim Sai* went on to *Mawar Sai* of Marcha and requested him to identify the reason for his son's bad condition. But also *Mawar Sai* replied helplessly that he had no knowledge about the disease and instead suggested *Rangchim Sai* to ask *Chung Sai* of his own village, Tidang.
>
>> [In this manner *Rangchim Sai* moved from the uppermost to the lowermost part of the Darma valley and asked the respective deities for help: Dhakar (*Layrato Sai*), Go [Hula] (*Dumphu*), Philam (*Narsati Sai*) Bon (*Gwala Sai*), Gankhu (*Nevangadang Sai*), Baling (*Punyu Sai*), Nangling (*Mudaru Sai*), Chal (*Pumo Sai*), Sela (*Puktang Sai*), Tingtha (*Nyungdang Sai*), Bungling (*Mongsaphu Sai*), Dar (*Yarsar Sai*), Sobla (*Hridyol Sai*), Guruguwa (*Patoto Sai*). But none of them knew what to do. On his way back to Tidang *Rangchim Sai* met a person called *Muktangajir*, who was able to see the future in his dreams. But he too was of no help. Full of disappointment *Rangchim Sai* passed through the villages of Son and Dugtu. Since he had previously taken the route on the opposite side of the Darma River, he had not visited those villages before. Without much hope he inquired again.]
>
> "Oh gods and goddesses, *Sajir* and *Mujir*, I bow down and touch your feet. Take my offerings and explain to me why my son is suffering. He neither eats food nor does he drink water. Please, tell me what to do!" *Mujir* and *Sajir* replied: "Oh *Rangchim Sai*, your beloved son is suffering from leprosy. Go and do something about it quickly! You may lose your beloved son!" Having heard this, *Rangchim Sai* started to lament: "I have received my son after so many difficulties and now he has caught such a terrible disease. How can I share this news with anybody?" He went home and opened the nine doors of a little bamboo box. He locked his beloved son in this box, so that nobody would learn about his condition.
>
> After many days, months and even a year had passed, the fourteen gods and thirty goddesses came together at the place named Helipang Danthu. They called all living creatures together and asked them to look for and bring back the beloved son of *Rangchim Sai*.

[69] At *Rangchim Sai's* shrines there are always two *alams*, which are installed some meters apart from each other. These are said to stand for his male and female part.

All animals and human creatures searched in the mountains, rivers and forests but they could not find him anywhere. Finally they visited *Rangchim Sai* and asked: “Nowadays we don’t see your son anymore. Where is he? Do you have any idea about his whereabouts?” But *Rangchim Sai* did not tell them the truth: “My son is not at home these days. He has gone far away to pursue his trade businesses”. With that answer all the human and animal creatures returned to the place called Helipang Danthu. But the gods and goddesses were wondering: “There is something wrong here! How could such an ill person like *Mangla Siri* have gone away to do business? We should call his friend *Dumphu Sai* from Hula village. He shall inquire again about the whereabouts of *Mangla Siri*”.

Dumphu Sai of Hula explained with his wise words: “Oh gods and goddesses, you don’t worry! I will find *Rangchim Sai’s* son and take him to a place from where this illness can never return to our golden land”. Hence Dumphu Sai packed his bag and moved away from Helipang Danthu. He passed the stream called Chyang Chyung, where all kinds of people used to gather. Then he arrived at the place called Bagbal Syeng Dar from where one reaches the top above Dhakar village. From there he followed the path that directly leads to *Rangchim Sai’s* residence. He asked *Rangchim Sai*: “Nowadays, my dear friend, your son has not been seen. Where has he gone? I am here to help. The fourteen great gods and the thirty great goddesses have agreed that I should take care of him. Do you keep him inside?” *Rangchim Sai* thought “How can I keep my beloved son inside that little box? It is better to hand him over to his best friend”. Thus *Rangchim Sai* opened the nine doors of the bamboo box and entrusted his son to *Dumphu Sai. Dumphu Sai* said: “You don’t worry! I will definitely cure your son. Just let us go!” He bid goodbye and left with *Rangchim Sai’s* son.

Dumphu Sai and *Mangla Siri* crossed the Darma River and arrived at the place of *Baigar Sai* named Likangdang. There *Dumphu Sai* said: “The road here remains closed in the month of August. We have to make an offering to *Baigar Sai* so that there will not be any accident on our way”. Then *Dumphu Sai* and *Rangchim Sai’s* ill son moved ahead and reached a waterfall named Gauto. *Dumphu Sai* said to his friend: “This water does not come down during the cold winter. We have to drink some water before moving ahead”. Then *Dumphu Sai* and *Mangla Siri* hurried on and arrived at a place where a lot of wild cumin was growing. *Dumphu Sai* said to his friend: “In this place people offer food and liquor prepared in their homes to the yaks. Let us do so as well!” Then they passed Thanve Dar and arrived at a wide, open place named Patthar in Rama Rautang. *Dumphu Sai* said: “In this place people used to adorn the horns of their yaks with ornaments of gold and silver. They also pull woolen covers around the animals’ feet and mark their back with a color before feeding them with good dishes and alcohol. Let us also take this opportunity!” When this was done they moved ahead and reached a place called Kal Punu.

Dumphu Sai said to his friend: “When people pass this place they usually pull a rope. If you pull the rope without a knot, you cannot be well but if you pull the rope with a knot, it will be possible for you to be cured”. When *Mangla Siri* saw three ropes in *Dumphu Sai’s* hand, he pulled one that had a knot. *Rangchim’s* son was stunned and said: “What happened? Isn’t my life ending? Will I be well or not?” *Dumphu Sai* replied: “You don’t worry let’s move ahead without delay!” *Dumphu Sai* and *Mangla Siri* passed Ganga Banga and then arrived at Bidang. At the big stone where travellers take a short rest *Dumphu Sai* said: “My dear friend, a difficult pass lies in front of us. When crossing this pass we need to consume what our forefathers preferred. Otherwise we will not succeed in crossing Sinla Dura”. When the two friends reached the mountaintop of Sinla, *Dumphu Sai* said: “It is too cold here. The frozen wind is covering us in ice. Let us eat the dough made of toasted flour [*syildu*] and let us drink some liquor!” Only then did the two friends depart from Sinla, heading towards the Byans valley.

In Byans *Dumphu Sai* and *Mangla Siri* first arrived at a place called Jyolinkong, which they quickly passed towards Kuti village, in which there stands the oldest tree of this golden land. Then they moved towards Gunji, a village that is favoured by the gods and goddesses of this land. *Dumphu Sai* said: "Let us attach a *daja* at the village shrine before moving ahead!" Next *Dumphu Sai* and *Mangala Siri* reached Kar Kawa, from where they crossed the Kali River. They passed the villages of Changru and Tinkar. From there they came to the Nima monastery and took the path towards Nara La in Humla. On this path *Dumphu Sai* said: "Why don't we do a long jump, my friend?" *Dumphu Sai* dug a deep, nine feet wide and nine feet long pit and explained: "First you jump, and then it's my turn". They agreed upon it and *Mangla Siri* jumped across. At the first attempt, he made it. He continued to jump across eight times in this manner. But then *Mangla Siri* got tired and when he jumped for the ninth time he fell down into that deep pit. *Dumphu Sai* saw that his friend was not able to come out anymore and covered him with a huge stone. Having got rid of this leper, he started his way back to Helipang Danthu.

> [On the way back he again passed through the places of the outbound trip. At each site *Dumphu Sai* calls to mind what he did there with *Mangla Siri* until he arrived at *Rangchim Sai's* place in Tidang.]

Rangchim Sai of village Tidang wondered: "You come alone? Where is my son? Why is he delayed? Why is he not coming with you?" *Dumphu Sai* could not tell *Rangchim Sai* about the death of his son and said instead: "Your son, my friend, went for a hunt in the Nausidi hills near Rama. We ran out of food and I am here to collect something. You do not need to worry! I will return with *Mangla Siri* in a couple of days". Then *Dumphu Sai* moved on to the friendly and just place called Helipang Danthu.

***

The gods and goddesses thought as well that they couldn't tell *Rangchim Sai* about the death of his beloved son. In order to prevent him from mourning, all the gods and goddesses collected a million little pieces of gold and silver. From that they forged a duplicate of *Mangla Siri*. When the duplicate was completed, they were not yet satisfied. Therefore, the gods and goddesses of all four directions came along with *syir sying* [*Juniper indica*], wooden torches, uncooked rice and colorful flags. When they made their offerings to the beautiful image, it started moving. A bright light, like that of the sun, could be seen on its face and its body was glittering like gold. And all the gods and goddesses were proud and happy with their work.

The gods and goddesses asked *Mangla Siri*: "Can you tell us where you live? Where are you from? Who are you? What are you doing here? Can you please tell us?" *Mangala Siri*, with his mild voice, told them: "I am the single son of *Rangchim Sai* from Tidang village. I came here to worship the gods and goddesses". The gods and goddesses became delighted when they heard these words and requested *Dumphu Sai*: "Please take this boy with you! Bring him home to his father *Rangchim Sai* in Tidang". After *Dumphu Sai* had taken *Mangla Siri* to Tidang, he said to *Rangchim Sai*: "Through the power of all the gods and goddesses *Mangla Siri* has been cured. Now your beloved son shall never again become ill!" Both *Mangla Siri* and his father were indescribably happy, and *Rangchim Sai* announced: "My son has returned due to the great efforts of all the gods and goddesses. I will always remain in their debt".

Rangchim Sai looked at his beloved son. The more he looked at him, the more beautiful he appeared to be. Time passed and *Mangla Siri* started travelling around in order to search for a beautiful girl. He visited each and every village, but there was not a single girl that was a suitable match for him. Desperately he returned back home and complained to his father in a frustrated tone: "There is not a single daughter in this land who suits me. Where can I go to find a girl with whom I can enjoy singing and dancing?" *Rangchim Sai* advised him: "My dear son, if you do not like the beautiful girls in this land, why don't you visit heaven and have a look at the nine daughters of the sun?" According to his father's suggestion *Mangla Siri* visited the solar family. And when he arrived there, his face instantly filled with happiness.

Catching sight of this unknown and utterly handsome man, the eight youngest daughters of the sun hid themselves and started flirting quietly. But only when *Mangla Siri* caught sight of the eldest daughter, his body instantly changed color. The eldest daughter was named *Ni Sino*. She welcomed *Mangla Siri* with her sweet voice and asked him to sit down. *Mangla Siri* replied with an appropriate greeting and delivered his speech: "You might be expecting an excuse for my visit, but let me be honest. There is not a single girl in our golden land who suits me. I thought it would be better to spend my time with you". The nine daughters of the sun happily accepted this proposal and they enjoyed themselves with eating and drinking, dancing and singing. All the nine sisters felt deeply attracted toward *Mangla Siri*. But it was only the eldest daughter *Ni Sino* who truly attracted *Mangla Siri*.

Engaged in this kind of amusement with the nine daughters of the sun, months went by quickly. But *Mangla Siri* did not even once think about returning back home. He did not care about spending years among the nine beautiful daughters of the sun. Eight years passed full of joy, but at the beginning of the ninth year some displeasure arose amongst them. The siblings increasingly spoke ill about each other and quarrels grew day by day. Each of the nine sisters claimed to be more beautiful and superior to the others. *Mangla Siri* soon got angry and finally decided to return back home. But the eldest daughter of the sun looked after him wistfully. She fixed her gaze on Tidang village and after a while the land was affected by great heat. The fields dried out, and water ran short also in the neighbouring hills, forests, fields, and pastures. The eldest daughter of the sun brought about a drought in Tidang village.

Mangla Siri got very weak and tired due to the severe heat and called for his father: "Oh father, please help me! I need some water, please give me some water!" *Rangchim Sai*, with his wise words, said to *Mangla Siri*: "My son, there are two water sources on the lower side of the village, just go there and quench your thirst". When *Mangla Siri* went there, he found that both sources were completely dried out. Thus *Mangla Siri* again requested his father: "The springs have no water and I am still thirsty. Please, give me some water!" *Rangchim Sai* replied: "If there is no water in these sources, then just go to the streams on the upper side of the village. There is water that comes down from the glaciers!" When *Mangla Siri* reached these streams, he also found them dry and said: "Oh father, there is not even a single drop of water and I am so thirsty, please give me something to drink!" *Rangchim Sai* replied: "If there is no water in the streams, just go to the banks of the river". *Mangla Siri* went to the river as his father suggested, but the river carried no water. Desperately he said to his father: "Oh father, there is no water in the river anymore! I am so thirsty, please give me some water!"

Having listened to this time and again, *Rangchim Sai* became himself concerned and said: "Let us go to the upper mountains. There you can quench your thirst from water that has accumulated between huge rocks". Holding hands and providing support, *Rangchim Sai* brought his son to an area called Sim-Sim Ti. When they reached there *Mangla Siri* immediately filled his mouth with water. But it was too late. *Mangla Siri* died at the first sip. Seeing this, *Rangchim Sai* went mad with distress and grief. He could not decide what to do and where to go. "How and why did this happen?" *Rangchim Sai* cried. And he continued mourning: "Beloved son! Where have you gone? Why have you left me alone?" *Rangchim Sai* pulled off his cap, held his head in his hands and promised his son's revenge: "I will only find some rest when I have drunken the blood of the sun's eldest daughter!"

***

After three days had passed, *Rangchim Sai* prepared to leave his home. He took some food, water, weaponry and other necessary items. Filled with the desire to take revenge for his beloved son, he headed toward Tibet in order to meet the famous archer *Phaiya Chalaphu Sai*.

[A description of the route is given.]

Rangchim Sai from Tidang was committed to take vengeance and he requested *Phaiya Chalaphu Sai* for his help: "Only you can solve my sorrows, please listen to me! I am in big trouble. I am affected by sorrow and grief, because I have lost my single son. I came here to ask for your assistance". The archer *Phaiya Chalaphu Sai* queried: "What kind of trouble are you in? Who is so foolish as to put his eyes upon you and evoke your enmity?" *Rangchim Sai* explained the situation: "Oh you great archer! My single son died because of the sun's eldest daughter. I want to shoot an arrow right through her eye". While *Rangchim Sai* spoke, he became full of sorrow and started mourning: "Oh god! My stars have become weak and put me into trouble. Oh *Phaiya Chalaphu Sai*, please take your bow. I wish you would shoot that arrow right through the eye of *Ni Sino*, the sun's eldest daughter. I would always remain in your debt".

Phaiya Chalaphu Sai was surprised at that request: "Why have you come so far and to this poor person? You are the greatest archer in this land!" But *Rangchim Sai* replied: "Because of my son's death, I lost my confidence. I am not able to do this. This is why I came that far to ask for your assistance". The archer *Phaiya Chalaphu Sai* was moved by *Rangchim Sai's* speech and took up his bow and arrow, put his feet firmly on the ground and looked toward the sky. He targeted the eye of *Ni Sino* and shot. But for five days there was not a single sign of the arrow. The archer assumed that he must have missed his target and told *Rangchim Sai*: "The arrow I shot must have gone lost". After this mishap, *Phaiya Chalaphu Sai* declared that he would not shoot another arrow. Out of embarrassment he injured the fingers of both hands, went away and hid.

Rangchim Sai raised his head toward the sky and doubted that he could ever be able to take revenge for the death of his son. He began to tremble with rage. "How could I ever take revenge?" he asked himself. The morning already dawned, and *Rangchim Sai* shed many tears. In great sorrow he embarked on his return journey. When he had crossed the pass into the Darma valley, the memory of his son's dreadful death entered his body like an attack of colic. His hands and feet were shivering with rage and *Rangchim Sai* grabbed his bow and an arrow. Without even looking he shot one arrow after the other into the sky. Only after having shot his ninth arrow, he saw that they all had hit right into an eye of *Ni Sino*.

Ni Sino did not know what had happened to her and yelled: "A clump of grass is stuck in my eye. Please, can someone come and take it out?" *Ni Sino*, screaming in pain called for her sisters who carefully examined her eye. They said: "This is not grass! Nine arrows hit you! These must have been shot by a man from earth!" When they found out that *Rangchim Sai* had taken revenge for the loss of his beloved son, *Ni Sino* got very angry. She was unhappy and thought that she looked pretty ugly now. She saw no value to live anymore. In her pain she decided to go underground and hide herself in a huge black gorge. Thus it became dark in the whole Darma valley, so dark that the light of the day completely disappeared.

Seasons passed, but there was still no sign of light. All the creatures went mad and the fourteen gods and the thirty goddesses sent a message to call all of them. When animal and human creatures arrived at Helipang Danthu the gods and goddesses said: "We need suggestions! How can we bring this situation to an end? Days have turned into nights and nights remain nights. We have never seen this before". *Layrato Sai* of Dhakar said: "Oh gods! All this is not a great surprise. *Rangchim Sai* from village Tidang shot nine arrows into the eye of *Ni Sino*, the eldest daughter of the sun. He did this to take revenge for his beloved son's death. When *Ni Sino* was not able to open one of her eyes anymore, she also closed her second one. Then she went underground and hid herself in a huge and dark cave. Since then she has disappeared from the world. There will be no chance to get the light back into our land unless she smiles again. We have to take the right decision in order to achieve this".

Puktan Sai of Go village said: "The gods and goddesses should ask the mouse for help. This small creature will be the right one to do this!" The fourteen great gods and the thirty great goddesses bowed their heads in front of the little mouse and said: "You are the only one who can do this. Search *Ni Sino* and make her smile again so that the sunlight will come back to this golden land once again!" Accepting the respectful greeting from the fourteen great gods and the thirty great goddesses, the mouse said: "You gods and goddesses, you need to prepare some food, a protecting cloth made of birch bark and a sword made of *sebali* [a tree species; unknown botanical name]. Provide these things to me and I will go and find *Ni Sino*. I will make her smile again so that the sunlight will come back to this golden land". The fourteen gods and the thirty goddesses prepared all these materials and gave it to the mouse before she departed from Helipang Danthu. The mouse walked to the west and after nine days she finally reached the dark place in which the eldest daughter of the sun was hiding herself. She spotted *Ni Sino* seated at a loom with closed eyes.

Seeing *Ni Sino's* condition, the mouse asked: "Oh eldest daughter of the sun! Why are you staying here in such great discomfort? Why is your hair open, why are so many tears running out of your eyes? Why have you taken off your earrings and jewellery? And why are you wearing your shawl back to front? Has anything bad happened in your home? Why are you looking so worried? Please tell me!" The eldest daughter of the sun told the mouse with her sweet voice: "Something bad has occurred in *Rangchim Sai's* house. That's why I have left my family, why I have suffered from hunger and thirst, and why I have shed so many tears. It's all because of *Rangchim Sai*. I am the only black sheep within the flock and I am the only black stripe on the face of a beautiful goat because of *Rangchim Sai*. I am the black stitching on a white woolen shawl and I am the black feather on a colorful bird because of *Rangchim Sai*. I am the black cloud on a blue sky and I am the dark spot on the bright moon for my desire that *Rangchim Sai* die. Everything I do is because of *Rangchim Sai's* wickedness".

Listening to the eldest daughter of the sun the mouse started to dance on the carpet. Poor mouse! It was a hilarious sight! She was wearing her strange clothing. She was showing her curious facial expressions. She was jumping and running up and down, clumsily wielding her sword. And pretty soon she was sweating heavily. *Ni Sino* shouted: "Oh innocent mouse, I warn you! Stop jumping and running around like that. Just go away from here! Otherwise I will kill you right away with my weaver's shuttle". But the mouse continued dancing even more wildly and with even more gestures than before. Her dress made of birch bark felt apart piece by piece, the sword broke into two pieces and she was completely soaked in sweat. At that moment *Ni Sino* lost control, opened her eyes and burst out in laughter. And as soon as she began laughing, the light of the day came back to the golden land of the Darma valley. Since *Ni Sino* was laughing and laughing, the mouse became happy too and returned back to Helipang Danthu.

When all the great gods and all the great goddesses thanked the little mouse, another difficulty arose. Since *Ni Sino* did not stop laughing, there was only day in the golden land of Darma. Once again all creatures met in Helipang Danthu. The gods and goddesses said: "What kind of situation is this? Now there are only days in this golden land? We have to decide about the proper length of the day and the length of the night!"

First of all, the large animals spoke out: "If there is night, it should be nine, and if there is day, it should be one". Similarly, the small reptiles and insects suggested that "If there is day, it should be nine, and if there is night, it should be one". But the gods and goddesses with their wise words said: "None of these suggestions are appropriate and just". Then it was the birds' turn. While the cunning crow was still busy calculating, the golden peacock raised his voice: "If there is day, it should be one, and if there is night, it also should be one". All the gods and goddesses instantly agreed: "This suggestion seems appropriate. All creatures will get justice from today onwards".

> After these wise words were spoken, all the creatures happily left the place. But when the peacock was about to leave the cunning crow approached him saying: "Brother, all my congratulations! But please listen to a few words!" Then, he continued with a mischievous smile: "You are the pride of all creatures! You proposed that fine solution. You are the most intelligent and just! I am highly inspired by you!" The little peacock became very cheerful and appreciated his fortune as well. They greeted each other in a friendly manner before the cunning crow continued: "Respected friend, in order to make our relation stable, we should dip our beaks into the glacial lake at night". Thus they set off to a lake in the upper mountains. But the cunning crow only dipped a beak of wood into the icy water, while the peacock did the same with his real beak. Suddenly, the golden peacock was helplessly fluttering around. He could not speak, as his beak was frozen. Then the cunning crow triumphantly said: "From your tongue came the suggestion of one day and one night. But I should have said this. Now I will cut out your tongue". In this way the cunning crow mercilessly cut out the tongue of the golden peacock.

Even to a non-affected outsider the narrative flow can trigger emotions and it is easy to imagine that such emotions are all the stronger for the bereaved. But how does this verbal performance allow the bereaved to deal with the painful death of a beloved family member? Bloch and Parry's (1982) influential approach to the phenomenon of mortuary rites helps to answer this question. Their approach builds upon a critical reading of Hertz's (1960) study 'Death and the Right Hand'. Hertz argues that traditional society is under constant threat because of the mortality of its individual members. Their death calls into question the authority and timelessness of the social order. Based on Durkheim's classical distinction between the individual and society, he then proposes that humans die not only as biological organisms but also as social persons on whom society has imprinted its identity. This collective identity is rescued through mortuary rites and subsequently recycled within the group, so that it becomes possible for the social body not only to survive death but also to thrive as a result of it. While Bloch and Parry (1982: 5) are also concerned with this "relationship between the biological individual and the social collectivity", they do away with the idea that society describes a kind of 'super-organism' (Ingold 2011: 90), or a reified entity that exists and acts for itself.

They argue instead that the occurrence of death provides people with an opportunity for "*creating* that 'society' as an apparently external force" (Bloch and Parry 1982: 6). In other words, mortuary rites facilitate the re-establishment and maintenance of social authority and order such as royal power and strength in a monarchical system. In that manner these practices can also deliver a solution to genuine political problems. According to Bloch and Parry this is indicated by the fact that death rituals are often charged with symbols of fertility and driven by processes of renewal. In this regard the authors make a significant observation: "[i]n most cases what would seem to be revitalized in funerary practices is that resource which is *culturally conceived* to be most essential to the reproduction of the social order" (Bloch and Parry 1982: 7; italics in original). Does this observation help to explain *sarat* and, more specifically, the mythological narratives that are chanted?

When the *amricha* commences his recitation, he renders inoperative the rules of conduct to which people are said to comply since time immemorial. In the course of the funeral rites the bereaved break these rules in order to facilitate a journey that runs contrary to the currents of life and the trajectories of becoming. Unlike the rituals described in the previous parts of this chapter, the flow of the river, the cycle of the seasons, the growth of animals and plants as well as the presence of the deities are no longer of concern. This becomes particularly obvious if one looks at the former *gvan*-ceremony. *Gvan* was only celebrated in the upper settlements, namely either in early spring or in late autumn. In other words, the journey of the soul was accomplished at liminal times when the villagers would themselves not have been able, or willing, to travel across the snowbound high passes; when the worshipping of the gods had either not yet started or had already passed; when work in the fields was not yet done or already done with; and when the migratory groups had either just arrived in or were about to leave their upper settlements.[70]

The yak, a highly valued livestock resource, that is – as we have seen before – important in the pastoral mobility system and ritually associated with the maintenance and reproduction of the social order, also played a crucial role in the former *gvan* ceremony. However, the manner in which the animal was treated in *gvan* seems to differ slightly from the model envisaged by Bloch and Parry. The yak was not revitalized but rather permanently expelled from the "field of relations whose unfolding is tantamount to the process of life itself" (Ingold 2000: 108): it was either chased away, or killed, or slaughtered and consumed by 'outsiders'. In the following I show that throughout the ceremony the animal remained central for ritually inverting the relations and flows of material exchange between wife-givers and wife-takers. In other words, the field of relations in which that livestock resource was integrated differed profoundly from that of ordinary life and social interaction. I further argue that even without the presence of the yak these inversions are still characteristic of *sarat*. They are key to understanding the relevance of *Mangla Syiri's* story as well as the cultural logic that drives the journey of the soul to the land of the dead. In order to make this more comprehensible, a few words about the story's main characters are due.

As mentioned before, the Darmani worship *Rangchim Sai* as an androgynous deity who personifies both parents of *Mangla Syiri*, i.e. of the deceased. Then there is *Dumphu Sai*, who accompanies *Mangla Syiri* on his way to death and, subsequently, delivers a duplicate of him to *Rangchim Sai*. Similar to the son-in-law during *sarat* he thus acts as a kind of facilitator. And lastly there are the sun's nine daughters, seductive lovers who are visited by *Mangla Syiri's* duplicate in search for a marriage partner. For reasons I explicate in the following, it is particularly the actions of these beautiful girls who shed some light on what the deceased person's 'sisters' do in *sarat* and why it was that the villagers got rid of the yak in the former *gvan*-ceremony. An article of clothing, which is typically associated with newly married girls, is a good entry point for analyzing these interrelations.

After the marriage ceremony a bride is supposed to wear a woolen scarf when visiting her in-laws' house for the first time. The garment is generally thought to emphasize the girl's decency. If she visits without it, this is not only understood as poor behavior but also as an offence against her parents-in-law. During *sarat*, however, the close female relatives or 'sisters' of the deceased – both married and

[70] If someone died in the summer settlement his or her body was often stored in a wooden box filled with salt until the traders returned from Tibet in September (Leder 2003: 110, footnote 78).

unmarried – wear such a scarf back to front. In my view this practice is indicative of another dimension that is stressed extensively by Bloch and Parry (1982: 18), namely the ritually established nexus between "death, female sexuality, human reproduction and natural fertility". In the textual body of *amrhimo* this nexus evolves in the relationship between the sun's nine daughters and *Mangla Syiri*: from the very beginning their relationship is fuelled by sexual desires and the boy fails to steer them towards a productive track. After years of enjoyment the situation finally creates jealousy and envy amongst the siblings. When *Mangla Syiri* runs away and returns back home without a bride, *Ni Sino* is disappointed and looks after him with wistful eyes. Her yearning gaze causes not only the land to dry out, but also her lover to die in the arms of *Rangchim Sai*.

In my interpretation the deceased person's 'sisters' (*chame*) – or of her husband, if the deceased is female – embody this uncontrolled, dangerous and unproductive power of female sexuality during *sarat*. Their wearing of the scarf upside down indicates this visibly. The son-in-law, a wife-taker to the deceased, not only guides this group of mourners but is also supposed to make all kinds of contributions during the ritual. As I have already touched upon in the previous chapter, the Darmani commonly practiced (matrilateral) cross-cousin marriage (Leder 2003: 73–81; cf. Allen 1975). According to the 'normal' pattern of interaction between exogamous lineages the flow of both women and material resources is directed from wife-giver to wife-taker: ego's maternal uncle, who is the father of ego's bride when following the pattern of (matrilateral) cross-cousin marriage, plays a special role in that regard. He shares, for instance, the greatest financial burden for the marriage ceremony of his daughter to ego. Ego assumes himself a similar role once the marriage of his daughter to his sister's son is about to happen. It is this transactional scheme between wife-givers and wife-takers, as it stretches across three generations, which is inverted during *sarat*: from the deceased person's son-in-law via his cognates to his maternal uncles.

The mourning group referred to as *mya-chame* consists of both the male in-laws and the close female kin of the deceased. This group was formerly held responsible for procuring the yak. The animal, which carried both the clothes and the bone of the deceased, was led around in the village by the son-in-law. On three consecutive days he brought it to every household in the village. The deceased person's 'sisters' fed it with plenty of liquor, beer and cooked food, and they would have shed their tears on the soon stupefied animal whilst bidding farewell. Towards the evening of each day the animal was then tugged to a spot somewhat away from the village center, where the kinsmen of the deceased pelt it with toasted grains. These were to nourish the newly departed soul prior to its long, difficult and dangerous journey to the land of the dead.

After the completion of that journey, the son-in-law again led the animal out of the village, this time, however, undecorated, where it was reviled, beaten and driven away by the male villagers. Both Sherring (1905: 114–115) and Allen (1974: 13) interpret this violent procedure as a final and most serious step that was required to get rid of the potentially dangerous soul. This interpretation is, however, somewhat imprecise, as in my view it was primarily neither the soul nor the animal toward which this violent act was targeted: the male in-laws of the deceased had sponsored the animal and one of them, typically the son-in-law, was responsible for it. The deceased person's 'sisters' welcomed and entertained the yak, like the sun's nine daughters are told to have done with *Mangla Syiri*. By doing so these women transferred the deadly powers of their reproductive capacities onto the animal's body. In my view it was these powers, through which *Ni Sino* is told to have caused the death of *Mangla Syiri*, which the male villagers then violently chased away toward the end of *gvan*.

When *Rangchim Sai* takes revenge on *Ni Sino* by shooting nine arrows into her eye, she feels ugly, goes underground and mourns. This causes prolonged darkness and infertility on earth. During the recitation of *amrhimo*, which always happens at night, the 'sisters' of the deceased voluntarily make themselves unattractive as well. They take off their jewelry, wear their hair loose and weep. What happens on the following morning provides a second reason for why the yak was formerly repelled from the spheres of daily social life: the male kin of the deceased de-install the shrine or *yan* and dispose of everything, except the deceased person's bone (*chye*), in a gorge nearby the village.[71] People say they have to make sure that there remains nothing left which attracts the departed soul and might allow it to re-enter into daily village affairs. The enduring presence of the yak would have made this possible. Following the disposal the male kin of the deceased shave their heads, whilst the 'sisters' get all dressed up before bringing at least one sheep or goat. These animals are slaughtered during the final part of the ceremony. Once that part is completed the maternal uncles of the deceased person receive a smaller – rather symbolic – portion of the inheritance as a recompense for the successful course of events.[72]

By this means the inverted transaction between wife-givers and wife-takers is completed: the son-in-law, together with the deceased person's sisters, sponsored the animals and various other material resources needed for the ceremony. The cognates, in turn, remunerate the maternal uncles of the deceased for the successful outcome of *sarat*. This outcome is celebrated in the early afternoon, until when the house has been thoroughly cleaned and the in-laws have prepared the food. All villagers come together and perform what is known as *sai phamo* ('god-speech'), i.e. the final branch of *amrhimo*. During *sai phamo* the *amricha* announces that the whole household has been established anew, that people are once more in proper relations, that the time of sorrow is over, that the grain bin is full of corn, that flocks of sheep and goats are available, that food is prepared and that all people have come together to worship the gods and goddesses and have a joint meal.

In former times all men and women would have concluded *sai phamo* with a dance throughout the whole village. People say that this used to be an amusing dance, which was meant to make everybody laugh and which turned their sorrow

[71] In earlier times the traders would have carried the bone with them to Tibet, where they buried it underground and covered it with piles of stones. Since the high passes to Tibet are closed these shrines are nowadays established in the uppermost part of the Darma valley, namely along the footpath connecting Tidang village with the camping ground called Bidang.

[72] The maternal uncles are ritually identified with a special term that is not used in common parlance, namely *puwahiya* (Sherring 1905: 113; Allen 1974: 92).

into happiness. The dance is only very rarely done in *sarat*. People say that with the decrease in seasonal migration the number of participants is normally too small in size to make this dance enjoyable. Instead the *amrhicha* only announces that the dead soul has left and that the ceremony was successful before the musicians start drumming in their ordinary rhythm patterns again. Whilst drinking, laughing and joking, people would slip banknotes into the musicians' caps and gradually get back into the 'ordinary' pattern of their day-to-day activities.

What I have left out of the analysis so far is the very beginning of *Mangla Syiri's* story. It all starts with the boy becoming ill. In search of help, his father *Rangchim Sai* travels southwards, to the lower part of the Darma valley. However, none of the gods and goddesses he meets on his way is able or willing to help. It is only on the way back that he comes to know that *Mangla Syiri* suffers from leprosy. In local understandings this contagious disease, like smallpox, is considered a threat to the ongoing process of social life, causing what Bloch and Parry (1982: 16) classify as a 'bad death', i.e. a death that represents "not merely a lost potential for, but is an actual threat to, fertility". According to Sherring (1905: 111) the Bhotiyas had such a 'horror' of leprosy "that if the deceased has been a victim of this dread disease, they simply cast the corpse into the water and retain nothing, not even a tooth".

Having learnt about the condition of his son, *Rangchim Sai* locks his son into a bamboo box until *Dumphu Sai* arrives for help. The two friends move away via the Byans valley toward Humla. Somewhere on the way, right in the middle of nowhere, *Dumphu Sai* tricks *Mangla Syiri* in a game with the result of *Mangla Syiri* loosing his life. But the gods and goddesses are not able to tell *Rangchim Sai* what has really happened. They therefore decide to manufacture a duplicate of *Mangla Syiri*. In a similar fashion, the son-in-law was formerly responsible to prepare an effigy of the deceased that was installed besides the soul's shrine (*yan*) during *amrhimo*. The only difference is that *Mangla Syiri's* effigy is not crafted of birch branches, but of precious metals. When this new body comes to life, the scene is set for the boy's encounter with the sun's nine beautiful daughters. Although this encounter ends with death as well, it at least gives *Rangchim Sai* the opportunity to identify and to take revenge on the forces that took away the life of his beloved son.

The association between leprosy, the sun and (in-)fertility is common in North India (Parry 1982: 104). Nevertheless, the manner in which people come to know about the way in which *Rangchim Sai* has dealt with the overall situation follows a pattern that features prominently in the mortuary rites of many groups speaking languages of the Tibeto-Burman family: a verbal journey. A common feature of these verbal journeys is that they do not aim to arrange the concrete physical environment according to an ideal form. Instead, they take place in the unadorned and familiar landscape. As András Höfer (1999: 229) comments, "[f]ar from finding the 'thusness' of the world meaningless, they [the Tibeto-Burman speaking groups] content themselves with namings and movements to mark what is invisibly immanent in the visible, real landscape".

The bereaved deal with death by narrating – and therewith making present – the journeys conducted by divine beings, humans, animals, celestial bodies and ultimately the dead soul itself. In these journeys neither the *amricha* nor his audience are actually required to move physically. Movements are rather accomplished through enumerations of place names, which typically present such a high degree of geographical precision that, as Oppitz (1999: 189) remarks with reference to the Magar in Nepal, "one could draw a reliable trekking map with them". However, it is not the point-to-point movement of objects *across* a ready-made surface, as represented on a map, which really matters. What rather is of concern in these journeys is that they proceed by means of telling stories and naming places, which is different from classifying objects and mapping space. In other words, people come to know the meaning of death by travelling through a 'storied world' (Ingold 2011: 141–143). As Ingold (2009: 203) puts it:

> Those who know well are able to *tell*. They can tell not only in the sense of being able to recount the stories of the world, but also in the sense of having a finely tuned perceptual awareness of their surroundings. Thus knowing *is* relating the world around you, and the better you know, the greater is the clarity and depth of your perception. To tell, in short, is not to represent the world but to trace a path through it that others can follow. Of course anthropologists have long recognized the educative functions of storytelling among people the world over. But they have been wrong to treat stories as vehicles for the intergenerational transmission of encoded messages that, once deciphered, would reveal an all-embracing system of conceptual categories. Stories do not, as a rule, come with their meanings already attached, nor do they mean the same for different people. What they mean is rather something that listeners have to discover for themselves, by placing them in the context of their own life histories".

The anthropological position that Ingold refers to and critiques in this quotation is the Durkheimian one. Durkheim separated the sphere of people's mutual engagement with each other through ritual, song and myth from that of their practical-technical interaction with environmental resources. For Ingold, by contrast, both sets of activities are primarily ways of dwelling in that they allow people to incorporate the fluctuations of the environment into the current of their taskscape. The aforementioned rituals for *Gabla Sai*, *Syang Sai* and *Su*, for instance, have highlighted that such practices do allow people not only to transform the world but also, and perhaps even more importantly, to participate "in the world's transformation of itself" (Ingold 2011: 8). When performing *sarat*, however, there is one major difference, namely that people deal with the unavoidable occurrence of death by establishing patterns of relations of what one might call, following Law and Mol (2001: 616), 'conjoined alterity'.[73] People achieve an inversion of the ordinary patterns and rhythms of their direct involvement with each other, and it is through this inversion that the world of the living and the world of the dead become mutually exclusive whilst remaining, at least partially, connected.

When I wrote down and translated these oral texts in Dharchula with an *amricha* who had become a good friend of mine, he urged me to do the transcription away

[73] While Law and Mol introduce this term to analyze spatial relations that characterize technological developments, they derived it from writings of Bachelard (1964) about the phenomenon on death. One might be tempted to analyze these patterns in terms of Victor Turner's (1969) concept of a liminal phase. However, one characteristic of liminality is that there is no structure but only anti-structure, i.e. sameness or absolute equality. I argue, on the contrary, that in *sarat* people achieve a kind of inversion that is itself structured.

from the busy town centre situated on the banks of the Kali River. His precaution is all too understandable. The patterns of relations that are enacted during *sarat* and traced in *amrhimo* can be both destructive as well as creative. Done properly, in the right place and at the right time, they make it possible for the newly departed soul not only to reach the land of the dead, but also remain there. But if done in the wrong way, at the wrong time or at the wrong place, these practices may trigger off other, supposedly much more unfavorable and dangerous outcomes. The narratives of *amrhimo*, which I have considered in this chapter, capture the imagination of the bereaved. They give form to their feelings and bad grievances in terms of what Rundstrom (1993: 21) calls a 'process cartography': a sort of mapping or surveying that is not meant to flatten the world of experience onto a two-dimensional plane, the map, but that rather attempts to draw people's attention "to an ever more intense poetic involvement" with their surroundings (Ingold 2000: 56).

This poetic involvement is characterized – as we have seen – by an inversion of the relationships in which the ritual participants are otherwise situated. People ritually establish patterns of relations in which the maintenance of experiential unity, completeness and continuity depends on, and is an effect of, discontinuity. Drum-rhythms and dancing-steps are inverted; garments are worn back to front; earrings and jewelry are taken off; the son-in-law, a wife-taker to the deceased, takes every provision to facilitate a transaction that runs contrary to the ordinary flow of things; and the yak, the most valuable of all livestock resources, was chased away for good and, perhaps, even till death, in former times. A villager once explained that the performance of *sarat* is like throwing a tree into a river and then making it move not only against the current but also upwards against gravity.

The departed soul is made to develop a kind of indifference or even disgust toward the world of the living without really noticing it. People say that the soul is then supposed to look at its former relations with different eyes. It is this modulation of the soul's perspective on the world of the living that suspends the 'natural' laws and makes the apparently impossible possible: a movement that runs against the current and against gravity. The forces of thrust that people therefore release are, without doubt, dangerous. The *amricha*, the skilled wayfarer, does not try to minimize but rather to harness these dangers in leading the soul to the realm of the dead. The bereaved facilitate this movement in that they temporarily also dwell along the inverted 'lines' of their ordinary relationships (Ingold 2006). Through this poetic inversion people exclude death and bind themselves ever deeper "into the web of life" (Ingold 2011: 68).

4.4 Coda

In the market-oriented border region of today's Uttarakhand, communal resource regulations can be identified as a focal point of state-society interaction (Agrawal 2005a, b). *Van panchayats*, village-based councils for regulating the use of forests and grasslands, are one of the oldest examples and they are of lasting importance to migratory groups such as the Bhotiyas (Nüsser and Gerwin 2008). Dominant approaches analyze these formally approved arrangements in terms of rational decision-making and by means of quantifiable parameters such as area size or numbers of users (Naidu 2009; Baland et al. 2010). And undoubtedly, such quantitative data is useful for assessing local situations and pointing out crucial patterns of resource use and sharing. However, scholars increasingly see the need to move beyond prescriptive approaches and the mechanistic models that are derived from it (Jones and Boyd 2011; Agrawal and Chhatre 2006).

Although the total number of livestock owned by the Kumaoni Bhotiyas has drastically reduced over the last decades, sheep and goat rearing still is the largest and most profitable component of animal husbandry in the high valleys inhabited by them. The ongoing presence of these animals also contributes to the operating efficiency of the *van panchayats*, especially through grazing fees that are imposed on a growing number of outsiders' flocks. While people complain – with good reason – about long and complex bureaucratic procedures, these institutions provide a common platform for cooperation among government officials, NGO workers, and pastoral communities. Several examples discussed in this chapter show how these different actor groups have become active partners in shaping everyday forms of land use: from tailoring the management procedures for pastures and forests to regulating the collection and legal sale of medicinal plants.

But while policy-makers in Uttarakhand endorse pastoral practices as a tool for alleviating poverty and enhancing a mountain specific development path, insufficient attention has so far been paid to the local cultural factors that provide "a crucial source of support and creativity within livelihood strategies" (Forsyth and Michaud 2011: 15). In pastoral research these factors are typically assessed in terms of environmental adaptation and risk prevention: people construe a view *of* the world that culturally mediates their relationship to the environment and ideally reduces the risk of livelihood failure. According to this approach people shape the world according to a pre-conceived image or received cognitive schemata. Here, however, I argued with Ingold (2000) that humans know the world by engaging with or dwelling *in* it. Guided by past experiences and grounded in an ongoing perceptual involvement with their surroundings they work things out whilst moving along. And they do this as much in the performance of ritual and the recitation of myth as when grazing animals or tilling fields.

The starting point for analyzing this improvisatory character of pastoral mobility formed a ritual sequence that unfolds between the time of sowing and harvest in the Darma valley. These rituals are visibly connected through the establishment, renewal and removal of certain tree trunks, which are called *alam* and devoted to particular deities in the upper settlements. We have seen that the *alam*-rituals involve different forms of physical movement through which particularly the male participants become "caught up in a reciprocal and quite muscular dialogue with the material" (Ingold 2000: 342). In the course of this dialogue the contours, textures and features of the environment are incorporated into various facets of mobility that are embodied and practiced. This process of incorporation is intimately tied to a wide range of agro-pastoral work activities, such as tilling the fields, wool shearing and others. Furthermore, seasonal variations are itself ceremonially harnessed to co-ordinate different labor tasks, to constitute and re-negotiate fields of identity and power, as well as to channel emerging conflicts beyond the spheres of settled life and seasonal migration.

In these rituals it is not only forest but also livestock resources, particularly yaks, which figure prominently. During the times of the former trans-Himalayan trade, yaks were particularly valued for their ability to cope with severe snow and difficult terrain conditions. In the rituals considered here, yaks are associated with attributes of purity and strength. They are further distinguished from sheep and goats, which stand for wealth and prosperity, as well as from cattle, which are bound up with the hardships and quarrels of village life. Most important, however, is the fact that yaks are conceived as willful and wayward agents whose presence is vital for the reproduction of social order. Because of these character traits the yak also stood at the centre of the traditional mortuary ceremony, called *gvan*, where it acted as a fellow wayfarer of the departed soul on its way to the land of the dead.

Already under British colonial rule the mortuary rites fuelled numerous discussions, especially because the yak was ultimately killed during the ritual. The whole ceremony has undergone significant change since then. Sheep and goats substituted the yak and the whole event was eventually renamed as *sarat*. One key element, however, endured: the recitation of an elaborated oral text. When the ritual specialist recites this text he takes the bereaved, and ultimately the departed soul itself, on a journey. This journey is based not so much on a muscular but a poetic dialogue with the environment. Movements are accomplished through the art of storytelling and the enumeration of place names. A special feature of these movements is that they are actuated by various inversions, particularly with regard to the interactions between wife-takers and wife-givers. Being immersed in the relational nexus of these inversions, people deal with death along their ordinary routes of travel but yet in a manner that is profoundly Other to the current of their ongoing livelihood activities.

Ingold has recently adopted the distinction between wayfaring and transportation to elucidate the distinction between various forms of human locomotion and the modes of apprehension they entail. The wayfarer *is* his movement because he sustains himself both perpetually and materially "through an active engagement with the country that opens up along his path" (Ingold 2011: 150). In transport, on the other hand, there is no need to incorporate into one's bodily movements the ever-shifting textures of the environment, since the traveler does not move but is moved through public means of traffic that make him "a passenger in his own body" (Ingold 2011: 150). Moreover, whereas the former has no final destination, the latter presupposes it as a foregone conclusion. And finally, whilst risks are calculated at the outset in transport, the wayfarer relies on his careful judgment and skill in coping with potentially unpredictable changes along his route. Whilst the ritual activities that comprise the *alam*-system clearly match what Ingold describes as wayfaring, the death rituals of the Darmani include both forms. During *sarat* people are immersed in the dwelt-in environment, but they modulate its force field in order to 'transport' the newly departed soul along the inverted 'lines' of their ordinary relationships into the beyond.

To conclude, it is through movement that pastoral groups like the Darmani make visible to themselves, and to others, their 'claims' to a world that is relentlessly changing (Rose 2012). Ritual and myth, as performative-poetic modes of dwelling, are integral to these claims, offering multiple constellations for incorporating the 'seen' and 'unseen' constituents of the environment into "rhythmic interrelations or patterns of resonance" (Ingold 2000: 154). These constituents include plants and animals, humans and deities, souls and spirits, weather and climate, as well as many other phenomena and forces that contribute to the world's becoming as it is sensed by those who dwell. But this process of incorporation never occurs in a vacuum, since the kinaesthetic and poetic aspects of mobility are always already politicized (Cresswell 2010). In this politics, care intermingles with exploitation, judgment with ignorance, skilled practice with mechanical operation, and resonance with opposition. However, it is by bringing qualitative questions of meaning, value and purpose to bear in our research that we can enrich our understanding of mountain pastoralists' mobility.

References

Agrawal A (2001) Common property theories and forest management in the Indian Himalaya. Contrib Indian Sociol 35(2):181–212

Agrawal A (2005a) Environmentality: technologies of government and the making of subjects. Duke University Press, Durham

Agrawal A (2005b) Environmentality: community, intimate government, and the making of environmental subjects in Kumaon, India. Curr Anthropol 46(2):161–190

Agrawal A (2005c) Dismantling the divide between indigenous and scientific knowledge. Dev Chang 26(3):413–439

Agrawal A, Chhatre A (2006) Explaining the success on the commons: community forest governance in the Indian Himalaya. World Dev 34(1):149–166

Agrawal A, Saberwal VK (2004) Wither South Asian pastoralism? An introduction. Nomadic Peoples 8(2):36–53

Allen NJ (1974) The ritual journey: a pattern underlying certain Nepalese rituals. In: von Fürer-Haimendorf C (ed) Contributions to the anthropology of Nepal. Proceedings of a symposium held at the School of Oriental and African Studies, University of London, June/July 1973. Aris and Philips, Warminster, pp 6–21

Allen NJ (1975) Byansi kinship terminology: a study in symmetry. Man New Ser 10(1):80–94

Axelby R (2007) 'It takes two hands to clap': how Gaddi shepherds in the Indian Himalayas negotiate access to grazing. J Agrar Chang 7(1):35–75

Bachelard G (1964) The psychoanalysis of fire. Beacon Press, Boston

Baland J-M, Bardhan P, Das S, Mookherjee D (2010) Forests to the people: decentralization and forest degradation in the Indian Himalayas. World Dev 38(11):1642–1656

Bauer KM (2004) High frontiers: Dolpo and the changing world of Himalayan pastoralists. Columbia University Press, New York

Bell C (1992) Ritual theory, ritual practice. Oxford University Press, New York

Bergmann C, Gerwin M, Sax WS, Nüsser M (2011) Politics of scale in a high mountain border region: being mobile among the Bhotiyas of the Kumaon Himalaya, India. Nomadic Peoples 15(2):104–129

Bergmann C, Gerwin M, Nüsser M, Sax WS (2012) State policy and local performance: pasture use and pastoral practices in the Kumaon Himalaya. In: Kreutzmann H (ed) Pastoral practices in High Asia: agency of 'development' effected by modernization, resettlement and transformation. Springer, Heidelberg, pp 175–194

Bickel B, Gaenszle M (1999) Introduction: cultural horizons and practices in Himalayan space. In: Bickel B, Gaenszle M (eds) Himalayan space: cultural horizons and practices. Völkerkundemuseum, Zürich, pp 9–32

Biersack A (1999) Introduction: from the 'new ecology' to the new ecologies. Am Anthropol 101(1):5–18

Bisht BS (2001) Ethnography of a tribe. Study of Anwals of Uttarakhand Himalaya. Rawat Publications, Jaipur

Blackburn S (2005) Die Reise der Seele. Bestattungsrituale und orale Texte in Arunachal Pradesh, Indien. In: Assmann J, Maciejewski F, Michaels A (eds) Der Abschied von den Toten: Trauerrituale im Kulturvergleich. Wallstein, Göttingen, pp 82–109

Blackburn S (2008) Himalayan tribal tales: oral tradition and culture in the Apatani valley. Brill, Leiden

Blaikie PM, Muldavin JSS (2004) Upstream, downstream, India, China: the politics of environment in the Himalayan region. Ann Assoc Am Geogr 94(3):520–548

Bloch M, Parry J (1982) Introduction: death and the regeneration of life. In: Bloch M, Parry J (eds) Death and the regeneration of life. Cambridge University Press, Cambridge, pp 1–44

Bourdieu P (1977) Outline of a theory of practice. Cambridge University Press, Cambridge

Coales O (1919) Economic notes on eastern Tibet. Geogr J 54(4):242–247

Cresswell T (2010) Towards a politics of mobility. Environ Plan D Soc Space 28(1):17–31

Dhakriyal DS (2004a) Himalayi Sauka sanskritik dharohar, vol I. Takshila Prakashan, New Delhi

Dhakriyal DS (2004b) Himalayi Sauka sanskritik dharohar, vol III. Takshila Prakashan, New Delhi

Douglas M (1966) Purity and danger. Routledge, London

Eckholm EP (1976) Losing ground: environmental stress and world food prospects. McLeod, Toronto

FES [Foundation for Ecological Security] (2004) Annual report 2003–2004. Foundation for Ecological Security, Anand

Forsyth T, Michaud J (2011) Rethinking the relationship between livelihoods and ethnicity in highland China, Vietnam and Laos. In: Michaud J, Forsyth T (eds) Moving mountains: ethnicity and livelihoods in highland China, Vietnam, and Laos. University of British Columbia Press, Vancouver, pp 1–27

Foucault M (1977) Discipline and punish. Allen Lane, London

Foucault M (1988) Technologies of the self. In: Luther MH, Gutman H, Hutton PH (eds) Technologies of the self: a seminar with Michel Foucault. Tavistock Publications, London, pp 16–49

Foucault M (1991) Governmentality. In: Burchell CG, Miller P (eds) The Foucault effect: studies in governmentality with two lectures by and an interview with Michel Foucault. Harvester Wheatsheaf, London, pp 87–104

Garbyal SS, Aggarwal KK, Babu CR (2005) Return of biodiversity in Darma Valley, Dharchula Himalayas, Uttaranchal, North India. Following fortuitous changes in traditional lifestyle of the local inhabitants. Curr Sci 88(5):722–725

Gerwin M, Bergmann C (2012) Geopolitical relations and regional restructuring: the case of the Kumaon Himalaya, India. Erdkunde 66(2):1–17

Gooch P (2009) Victims of conservation or rights as forest dwellers: Van Gujjar pastoralists between contesting codes of law. Conserv Soc 7(4):239–248

GoU [Government of Uttaranchal/Uttarakhand] (2001) The Uttaranchal panchayati forest rules – 2001. Forest Department, Dehra Dun

GoU [Government of Uttaranchal/Uttarakhand] (2003) Livestock survey for the Tehsil Munsiari, 2001 (unpublished document)

GoU [Government of Uttaranchal/Uttarakhand] (2004) Uttaranchal decentralised watershed development project: tribal and transhumant strategy. Watershed Management Directorate, Dehradun

GoU [Government of Uttaranchal/Uttarakhand] (2010a) Transhumant plan. Watershed Mangement Directorate, Veterinary Department for the District Pithoragarh, Dehradun

GoU [Government of Uttaranchal/Uttarakhand] (2010b) Government order 98-PA/8-3[2]. Forest Department of Uttarakhand, Dehradun

Guha R (1989) The unquiet woods. Ecological change and peasant resistance in the Himalaya. Oxford University Press, New Delhi
Guha R (2001) The prehistory of community forestry in India. Environ Hist 6(2):213–238
Guillet D (1983) Toward a cultural ecology of mountains: the central Andes and the Himalayas compared. Curr Anthropol 24(5):561–574
Harris M (1998) The rhythm of life on the Amazon floodplain: seasonality and sociality in a riverine village. J R Anthropol Inst 4(1):65–82
Heidegger M (1971) Poetry, language, thought [translated and introduction by A. Hofstadter]. Harper and Row, New York
Hertz R (1960) Death and the right hand [translated by R. and C. Needham, with an introduction be E.E. Evans-Pritchard]. Cohen and West, London
Hjort A (1982) A critique of 'ecological' models of pastoral land use. Nomadic Peoples 10:11–27
Höfer A (1999) *Nomen est numen*: notes on the verbal journey in some western Tamang oral ritual texts. In: Bickel B, Gaenszle M (eds) Himalayan space: cultural horizons and practices. Völkerkundemuseum, Zürich, pp 205–246
Ingold T (1980) Hunters, pastoralists and ranchers. Cambridge University Press, Cambridge
Ingold T (2000) The perception of the environment. Essays in livelihood, dwelling and skill. Routledge, London
Ingold T (2006) Rethinking the animate, re-animating thought. Ethnos 71(1):9–20
Ingold T (2009) Stories against classification: transport, wayfinding and the integration of knowledge. In: Bamford S, Leach J (eds) Kinship and beyond: the genealogical model reconsidered. Berghahn, New York, pp 193–213
Ingold T (2011) Being alive: essays on movement, knowledge and description. Routledge, London
Ingold T, Kurttila T (2000) Perceiving the environment in Finnish Lapland. Body Soc 6(3–4):183–196
Ives JD, Messerli B (1989) The Himalayan dilemma: reconciling development and conservation. Routledge, London
Jones L, Boyd E (2011) Exploring social barriers to adaptation: insights from western Nepal. Glob Environ Chang 21(4):1262–1274
Kapila K (2008) The measure of a tribe: the cultural politics of constitutional reclassification in north India. J R Anthropol Inst 14(1):117–134
KBPF [Kumaon Bhotiya Peoples' Federation] (1947) The report of the Kumaon Bhotiya Peoples' Federation. Submitted before the minority sub-committee of the constituent assembly's advisory committee on 17th April 1947
Krätli S, Schareika N (2010) Living off uncertainty: the intelligent animal production of dryland pastoralists. Eur J Dev Res 22(5):605–622
Kreutzmann H (2001) Development indicators for mountain regions. Mt Res Dev 21(2):132–139
Kreutzmann H (2006) People and mountains: perspectives on the human dimension of mountain development. Glob Env Res 10(1):49–61
Kreutzmann H (2012) Pastoral practices in transition: animal husbandry in high Asian contexts. In: Kreutzmann H (ed) Pastoral practices in High Asia: agency of 'development' effected by modernization, resettlement and transformation. Springer, Heidelberg, pp 1–29
Lamb A (1989) Tibet, China and India, 1914–1950: a history of imperial diplomacy. Roxford Books, Hertfordshire
Landor AH (1899) In the forbidden land: an account of a journey in Tibet, capture by the Tibetan authorities, imprisonment, torture, and ultimate release. William Heinemann, London
Law J, Mol A (2001) Situating technoscience: an inquiry into spatialities. Environ Plan D Soc Space 19(5):609–621
Leach M, Mearns R, Scoones I (1999) Environmental entitlements: dynamics and institutions in community-based natural resource management. World Dev 27(2):225–247
Leder S (2003) Zu einer Zeit, als Bäume und Gräser noch sprechen konnten: Sozioökonomie, Kosmologie und Mythologie der Rang-Shauka im zentralindischen Himalaya. Peter Lang, Frankfurt am Main

Mauss M (with Henri Beuchat) (1979) Seasonal variations among the Eskimo: a study in social morphology. Routledge, London
Mazullo N, Ingold T (2008) Being along: time and movement among the Sámi people. In: Bærenholdt J, Granås B (eds) Mobility and place: enacting northern European peripheries. Ashgate, Aldershot, pp 27–38
Messer E, Lambek M (eds) (2001) Ecology and the sacred: engaging the anthropology of Roy A. Rappaport. University of Michigan Press, Ann Arbor
Metz JJ (2010) Downward spiral? Interrogating narratives of environmental change in the Himalaya. In: Guneratne A (ed) Culture and the environment in the Himalaya. Routledge, London, pp 17–39
Mittal AK (1986) British administration in Kumaon Himalayas. Mittal Publications, New Delhi
Mohan CR (2007) Soft borders and cooperative frontiers: India's changing territorial diplomacy towards Pakistan and China. Strateg Anal 31(1):1–23
Montero RG, Mathieu J, Singh C (2009) Mountain pastoralism 1500–2000: an introduction. Nomadic Peoples 13(2):1–16
Murray WH (1951) Scottish Himalayan expedition, 1950. J.M. Dent & Sons Ltd., London
Naidu SC (2009) Heterogeneity and collective management: evidence from common forests in Himachal Pradesh, India. World Dev 37(3):676–686
Nawa K (2007) Some unintended consequences of ritual change: the case of funerals in Changru, Byans, far western Nepal. In: Ishii H, Gellner D, Nawa K (eds) Social dynamics in northern South Asia, vol 2: political and social transformations in north India and Nepal. Manohar, New Delhi, pp 263–288
Niamir-Fuller M (1999) Toward a synthesis of guidelines for legitimizing transhumance. In: Niamir-Fuller M (ed) Managing mobility in African rangelands. The legitimization of transhumance. Intermediate Technology Publications, London, pp 266–290
Nightingale AJ (2005) 'The experts taught us all we know': professionalization and knowledge in Nepalese community forestry. Antipode 37(3):581–604
Nüsser M (1998) Nanga Parbat (NW-Himalaya): Naturräumliche Ressourcenausstattung und humanökologische Gefügemuster der Landnutzung. Dümmler, Bonn
Nüsser M (2006) Ressourcennutzung und nachhaltige Entwicklung im Kumaon-Himalaya (Indien). Geogr Rundsch 58(10):14–22
Nüsser M, Gerwin M (2008) Diversity, complexity and dynamics: land use patterns in the central Himalayas of Kumaon, northern India. In: Löffler J, Stadelbauer J (eds) Diversity in mountain systems. Asgard-Verlag, Sankt Augustin, pp 107–119
Oppitz M (1999) Cardinal directions in Magar mythology. In: Bickel B, Gaenszle M (eds) Himalayan space: cultural horizons and practices. Völkerkundemuseum, Zürich, pp 167–204
Pant SD (1935) The social economy of the Himalayas. Based on a survey in the Kumaon Himalayas. George Allen and Unwin, London
Parry J (1982) Sacrificial death and the necrophagous ascetic. In: Bloch M, Parry J (eds) Death and the regeneration of life. Cambridge University Press, Cambridge, pp 74–110
Polit K (2012) Women of honour: gender and agency among Dalit women in the central Himalayas. Orient BlackSwan, New Delhi
Prasad RR (1989) Bhotia tribals of India. Dynamics of economic transformation. Gian Publishing House, New Delhi
Radcliffe-Brown AR (1922) The Andaman islanders: a study in social anthropology. Cambridge University Press, Cambridge
Rappaport RA (1967) Ritual regulation of environmental relations among a New Guinea people. Ethnology 6(1):17–30
Rappaport RA (1968) Pigs for the ancestors: ritual in the ecology of a New Guinea people. Yale University Press, New Haven
Rhoades RE, Thompson SI (1975) Adaptive strategies in alpine environments: beyond ecological particularism. Am Ethnol 2(3):535–551
Richards P (1993) Cultivation: knowledge or performance? In: Hobart M (ed) An anthropological critique of development: the growth of ignorance. Routledge, London, pp 61–78

Robbins P (2001) Tracking invasive land covers in India, or why our landscapes have never been modern. Ann Assoc Am Geogr 91(4):637–659
Rose M (2012) Dwelling as marking and claiming. Environ Plan D Soc Space 30(5):757–771
Rundstrom RA (1993) The role of ethics, mapping, and the meaning of place in relations between Indians and Whites in the United States. Cartographica Int J Geogr Inf Geovisualization 30(1):21–28
Saberwal VK (1999) Pastoral politics: shepherds, bureaucrats, and conservation in the western Himalaya. Oxford University Press, New Delhi
Sarin M (2001) Empowerment and disempowerment of forest women in Uttarakhand, India. Gender Technol Devel 5(3):341–364
Sax WS (1991) Mountain goddess: gender and politics in a Himalayan pilgrimage. Oxford University Press, New York
Sax WS (2010) Ritual and the problem of efficacy. In: Sax WS, Quack J, Weinhold J (eds) The problem of ritual efficacy. Oxford University Press, New York, pp 3–16
Scoones I (1995) New directions in pastoral development in Africa. In: Scoones I (ed) Living with uncertainty. New directions in pastoral development in Africa. Intermediate Technology Publications, London, pp 1–36
Scoones I (1999) New ecology and the social sciences. What prospects for a fruitful engagement? Ann Rev Anthropol 28:479–507
Sherring CA (1905) Notes on the Bhotias of Almora and British Garhwal. Regional Archive Uttarakhand, Nainital
Sherring CA (1974 [1916]) Western Tibet and the British Borderland. The sacred country of Hindus and Buddhists. With an account of the government, religion and customs of its peoples. Cosmo Publications, Delhi
Singh NM (2013) The affective labor of growing forests and the becoming of environmental subjects: rethinking environmentality in Odisha, India. Geoforum 47:189–198
Skaria A (2005) Comment to A. Agrawal's 'Environmentality: community, intimate government, and the making of environmental subjects in Kumaon, India'. Curr Anthropol 46(2):184–185
Springate-Baginski O, Blaikie P (eds) (2007) Forests, people and power: the political ecology of reform in South Asia. Earthscan, London
Stellrecht I (1992) Umweltwahrnehmung und vertikale Klassifikation im Hunza-Tal (Karakorum). Geogr Rundsch 44(7–8):426–434
Taylor M (2015) The political ecology of climate change adaptation: livelihoods, agrarian change and the conflicts of development. Routledge, London
Tolia RS (2010) Livestock rearing for poverty reduction. Garhwal Post, March 21
Townsend JG, Porter G, Mawdsley E (2004) Creating spaces of resistance: development NGOs and their clients in Ghana, India and Mexico. Antipode 36(5):871–889
Traill GW (1992 [1832]) Statistical report on the Bhotiya mehals of Kumaon. In: Joshi MP, Fanger AC, Brown CW (eds) Himalaya: past and present, vol 2. Shree Almora Book Depot, Almora, pp 99–154
Turner V (1969) The ritual process. Structure and anti-structure. Routledge, London
Wagner A (2013) The Gaddi beyond pastoralism: making place in the Indian Himalaya. Berghahn, New York
Winkler D (2008) *Yartsa gunbo* (*Cordyceps sinensis*) and the fungal commodification of Tibet's rural economy. Econ Bot 62(3):291–305
Winkler D (2010) *Cordyceps sinensis*: a precious parasitic fungus infecting Tibet. Field Mycol 11(2):60–68

Chapter 5
Epilogue

Abstract This chapter concludes the study on the Himalayan border region and summarizes its main arguments. Based on an analytical framework that is grounded in and contributes to recent advances in the field of border studies, I explored how the Bhotiyas have used their agency to develop a flourishing trans-Himalayan trade under British colonial influence in Kumaon; to assert an identity and win legal recognition as a tribal community in the political setup of independent India; and to innovate their pastoral mobility in the context of ongoing state and market reforms in present day Uttarakhand. Two directions for follow-up research are highlighted, namely, on the one hand, the changing translocal networks within which Himalayan communities seek to achieve viable livelihood outcomes and, on the other hand, the application of 'Zomia-thinking' to critically examine emergent science-policy domains such as the management of socio-ecological systems through a resilience-based framework.

Keywords Himalayan border region • Bhotiyas • Trans-Himalayan trade • Identity • Mobility • Agency • Translocal networks • Zomia • Resilience • Kumaon • Uttarakhand • India

In this book I have analyzed how the Bhotiyas, a Scheduled Tribe of India, have encountered and negotiated their positionality vis-à-vis a whole range of agents and forces that impinged on their daily lives and livelihoods in the historically evolving constellations of a 'high mountain border region' (Bergmann et al. 2008): from the officialdom of the British empire to the founding fathers of the Indian nation state; from international conflicts to inter- and intra-community rivalries; from seasonal fluctuations in the environment to new market developments; and from divine beings and evil spirits to the departed soul of a beloved family member. The study has been grounded in and contributes to recent advances in the growing field of border studies. It established an analytical framework that allowed me to, first, appreciate the centrality of this Himalayan border region within the entangled imperial and national histories of adjacent states, and, second, to explore how its residents, the Bhotiyas, used their agency to develop a flourishing trans-Himalayan trade under British colonial influence; to assert an identity and win legal recognition as a tribal community in the political setup of independent India; and to innovate

C. Bergmann, *The Himalayan Border Region*, Advances in Asian Human-Environmental Research, DOI 10.1007/978-3-319-29707-1_5

their pastoral mobility in the context of ongoing state and market reforms in present day Uttarakhand.

A point of departure for the study was the ongoing debate over 'Zomia', a neologism that is increasingly being used to conceive High Asia as a continuous zone and an agentive site of political action (van Schendel 2002; Scott 2009; Michaud 2010). Shneiderman has fruitfully applied 'Zomia-thinking' to analyze processes of state- and political subject-formation in the contact zones of Himalayan polities that remained independent throughout the colonial era, namely Nepal and Tibet (Shneiderman 2013). Although the histories and geographies of European colonization and frontier imperialism also play a role in this particular setting, the case of Kumaon provided the alternative analytical focus of a Non-regulation Province. Whilst such provinces formed part of the British Empire, they were administered not by the rule of law but rather "by discretion or executive interposition" (Misra 1977: 81). In Kumaon this legal status was finally dissolved in 1925. However, the region's marginalization extended into the post-colonial period until Uttarakhand was carved out as an independent state of the Indian Union in 2000.

Competing polities have at various times aimed to extend their influence into the high valleys inhabited by the region's trans-Himalayan traders, the Bhotiyas (Bergmann 2016: 11). Before the onset of British colonial rule in this area, the interactions between the ruling dynasties of Kumaon, Tibet and Nepal created a regionally arranged power geometry. Control over the trans-Himalayan trade was a focus of special attention for these rulers, both in order to regulate the movement of people and commodities, and to extract some kind of surplus from it. In that manner the traders were positioned in a complex geography of graded and partially overlapping shades of political, economic and cultural authority. Their trade routes connected different spheres (for example, settlement zones and polities) and phases (such as agricultural seasons and ritual cycles) of action. The emerging junctures of this regional *ecumene* – such as trade depots or high passes – opened up multiple means of articulating sovereign claims, strengthening territorial power and procuring goods. The British added a new force to the underlying negotiations over sovereignty and territoriality, when they conquered the Kumaon Himalaya in 1815. This campaign of conquest provided them with their first direct access route to the Tibetan plateau.

Van Schendel (2007: 37) argues that the explosive nature and continuing relevance of the Sino-Indian boundary dispute stems largely "from failed border-making exercises in the colonial period". Based on the results of Chap. 2, I could specify this assessment by arguing that the conflict has been nourished by the rather persistent presence of confluent territories and overlapping sovereignties during the nineteenth century. Overlapping sovereignties occur when different actors claim control over the same sort of thing, for instance trade. Whilst these claims are often imagined as being inherently territorial, their articulation can also, and at the same time, be marked by a great uncertainty as to who controls where. Without the presence of definite boundaries, actors can effectively negotiate their political authority within such confluent territories through power relationships of allegiance. In this context I have highlighted the plural array of processes and actor constellations that contributed to the manner in which the British articulated their imperial sovereignty in Kumaon. The region's frontier people, the Bhotiyas, were far from being passive victims or mere recipients of elite imperial plans. On the contrary, they actively negotiated and co-produced the forces through which the Himalayan border region evolved and shifted its shape as an imperial environment.

The question of how the Bhotiyas have capitalized on their intimate acquaintance with the workings of imperial power to assert who they are or want to be within the evolving configurations of an independent nation state was addressed in Chap. 3. The chapter analyzed the Bhotiyas struggle for tribal recognition in terms of a contentious and ongoing 'scalar politics' (MacKinnon 2011). In such a politics scale matters less as an external framework for action than as an 'effect' that actors produce through the practical and discursive strategies they deploy whilst pursuing their various agendas (Kaiser and Nikiforova 2008). In this regard I could demonstrate how *Bhot* (or the region inhabited by the Bhotiyas) was produced as a scale effect in independent India when influential community members staked a claim for legal recognition as a Scheduled Tribe (ST) in 1947; how, conversely, independent India was reproduced as a scale effect in *Bhot* when less privileged community members aimed to overcome the rather unequal power relations that had served to substantiate this claim after the Bhotiyas' public notification as a ST in 1967; and, at the time of Uttarakhand's independence in 2000, how caste Hindus went to court and promoted the idea that the Bhotiyas' ST-designation is a scalar expression that does not pertain to a specific *people* but to a *region*, including all 'native' residents. The analysis showed that whilst there existed a well-established and politically active 'tribal' elite even before India's independence, the ST-status also enabled wider sections of the community to take advantage of related government support schemes whilst, simultaneously, contesting the effect of scale through which their tribal identity had initially been established.

In recent years the Government of Uttarakhand started to promote, as it was explicated in Chap. 4, the 'tribal and transhumant strategy' of the Bhotiyas as a means to reduce poverty and foster sustainable livelihoods (GoU 2004: 5). However, whilst the mobile practices of agro-pastoral communities have received scholarly attention as a strategy that allow these people to make an efficient use of 'marginal resources' (Ehlers and Kreutzmann 2000), their rituals and myths have either been neglected or considered of only secondary importance in this connection. In order to counteract this tendency I argued with Ingold (2000) that humans know the world by engaging with or dwelling *in* it, which they do as much, although differently, in the performance of a ritual or the recitation of a myth as when grazing animals or tilling fields. On this conceptual basis I then explored two sets of ritual activity in the Darma valley, both of which involved various kinds of movement along the vertical gradient of the mountain environment. The chapter demonstrated that within the context of these activities the Darmani make visible to themselves, and to others, their lived experiences of and situated 'claims' to a world whose constituents are relentlessly changing (Rose 2012). These constituents include biotic and livestock resources, humans and deities, souls and spirits, weather and climate, as well as many other phenomena and forces that contribute to the world's becoming as it is sensed by those who dwell. By discussing the qualitative questions of meaning, value and purpose that such a ‘poetics of dwelling’ (Ingold 2000: 11) implies, the obtained results complement the quantitative metrics through which results-oriented development research tends to assess pastoral mobility and resource management strategies in terms of efficiency, optimization or sustainability.

Follow-up research is conceivable in (at least) two major directions: first, while my study has largely focused upon the Bhotiyas in the Kumaon division of Uttarakhand, the historical and contemporary ties that these people have maintained with other groups along the Indo-Tibetan interface are still poorly understood. Examples include their former tributary relations to polities in western Nepal as well as their contemporary employment of Gaddi shepherds from Himachal Pradesh. Such translocal networks, which span across both federal and national state borders, form an important asset with regard to the ways in which Himalayan communities seek to achieve viable livelihood outcomes and thus clearly deserve further scientific attention (Greiner and Sakdapolrak 2013; Benz 2014). Second, Zomia-thinking seems to be ideally suited to unsettle conventional resilience-thinking, particularly its tendency to make certain forms of self-organisation, hierarchy and development, mainly the neoliberal, dominate over alternative or even radically ‘other’ visions (Evans and Reid 2014). For Zomians a lack of access, poor infrastructure and harsh environmental conditions have not merely established hindrances but also provided opportunities. They could, for instance, engage in a lucrative cross-border trade without having to face great burdens in terms of administration, taxation and legal regulation. It remains an open question whether and how marginalized and oppositional communities’ in High Asia use their culturally embedded agency to engage with the human-induced environmental changes – including the logics, practices and discourses of related research and policy agendas – that increasingly affect their region.

References

Benz A (2014) Mobility, multilocality and translocal development: changing livelihoods in the Karakoram. Geogr Helv 69:259–270

Bergmann C (2016) Confluent territories and overlapping sovereignties: Britain’s nineteenth-century Indian empire in the Kumaon Himalaya. J Hist Geogr 51:88–98

Bergmann C, Gerwin M, Nüsser M, Sax WS (2008) Living in a high mountain border region: the case of the ‘Bhotiyas’ of the Indo-Chinese border region. J Mt Sci 5:209–217

Ehlers E, Kreutzmann H (eds) (2000) High mountain pastoralism in northern Pakistan. Steiner, Stuttgart

Evans B, Reid J (2014) Resilient life: the art of living dangerously. Polity Press, New York

GoU [Government of Uttaranchal/Uttarakhand] (2004) Uttaranchal decentralised watershed development project: tribal and transhumant strategy. Watershed Management Directorate, Dehradun

Greiner C, Sakdapolrak P (2013) Translocality: concepts, applications and emerging research perspectives. Geogr Compass 7(5):373–384

Ingold T (2000) The perception of the environment. Essays in livelihood, dwelling and skill. Routledge, London

Kaiser R, Nikiforova N (2008) The performativity of scale: the social construction of scale effects in Narva, Estonia. Environ Plan D Soc Space 26(3):537–562
MacKinnon D (2011) Reconstructing scale: towards a new scalar politics. Prog Hum Geogr 35(1):21–36
Michaud J (2010) Editorial – zomia and beyond. J Glob Hist 5(2):187–214
Misra BB (1977) The bureaucracy in India. An historical analysis of development up to 1947. Oxford University Press, New Delhi
Rose M (2012) Dwelling as marking and claiming. Environ Plan D Soc Space 30(5):757–771
Scott JC (2009) The art of not being governed. An anarchist history of upland Southeast Asia Yale University Press, New Haven
Shneiderman S (2013) Himalayan border citizens: sovereignty and mobility in the Nepal-Tibetan Autonomous Region (TAR) of China border zone. Polit Geogr 35:25–36
van Schendel W (2002) Geographies of knowing, geographies of ignorance: jumping scale in Southeast Asia. Environ Plan D Soc Space 20(6):647–668
van Schendel W (2007) The Wagah syndrom: territorial roots of contemporary violence in South Asia. In: Basu A, Roy S (eds) Violence and democracy in India. Seagull Books, Oxford, pp 36–82

Index

A

Adivasi. *See* Scheduled Tribe (ST)
Affirmative action, 4, 13, 82, 107, 108
Afghanistan, 43, 85
Africa, 128
Agency, 4, 5, 16, 27, 28, 62, 64, 83, 87, 88, 113, 128, 133, 185, 188
Agriculture, 142
  combined mountain, 4
Alam. *See* Ritual
Almora, 34, 35, 46, 47, 56–58, 91, 93, 98, 100, 103
Amrhimo, 163, 164, 173–175, 177
Amricha, 162–164, 171, 174–177
Anglo-Nepalese war. *See* War
Anthropometry, 12, 81
Anusuchit janjati. *See* Scheduled Tribe (ST)
Anwal, 9, 137
Askot, 60, 102
  Rajwar of Askot, 32, 56, 60, 157
Atis. *See* Medicinal plants
Auckland's Folly. *See* War
Autoethnographic text, 77

B

Bai. *See* Buckwheat
Banricha, 162, 164
Bhot, 7, 79, 87, 88, 92–107, 109, 111, 116–118, 187
Bhotiyas, Garhwal
  Jadh, 8
  Marcha, 8, 9
  Tolcha, 8
Bhotiyas, Kumaon
  Byansi, 8
  Chaudansi, 8
  Darmani, 8, 9
  Johari, 8
Bod. *See* Tibet
Borax, 8, 31, 45, 47, 58, 60
Border
  Border Area Development Program, 109, 142
  Border Bhotiya Uplift Committee, 108–111
  borderland, 4–6, 17, 29, 30, 38, 76, 85
  border studies, 5–7, 185
  Border Youth Welfare Association, 104, 115
  Himalayan border region, 4–7, 10, 16, 24, 26–42, 55, 62, 64, 185, 187
  Indo-Tibetan border, 9, 138, 153
  Sino-Indian border war (*see* War)
Boundary
  ethnic boundary, 85
  traditional customary boundary line, 61, 62
British
  British East India Company, 10, 23, 24, 39, 46
  British prestige, 77
Buckwheat
  Fagopyrum esculentum (*palti*), 9, 31, 141
  Fagopyrum tataricum (*bai*), 9, 31, 141
Buddhist, 7, 12, 13, 35, 36, 49, 50, 93, 103, 105, 116, 144, 163
  crypto-Buddhist, 15, 105
Bugyal. *See* Pasture
Byans valley, 8, 9, 16, 35, 44, 46, 47, 54, 56, 88, 98, 115, 116, 151, 166, 175

C. Bergmann, *The Himalayan Border Region*, Advances in Asian Human-Environmental Research, DOI 10.1007/978-3-319-29707-1

C
Carrying capacity, 128, 137
Central Asia, 12, 26, 40, 48, 78
Central Himalaya, 3, 32
Chand dynasty, 32
Chaudans valley, 8, 35, 56, 97, 98
China, 3, 6, 13, 26, 27, 29, 43, 44, 60, 62, 82, 88, 98, 128, 129, 136, 140
Cordyceps sinensis, 139–141
Cremation. *See* Ritual
Cyakti. *See* Liquor

D
Dalit, 3, 77
Darma valley, 2, 3, 8, 9, 15, 16, 25, 30, 31, 33, 35, 37, 78, 87–89, 97, 101, 103, 104, 115, 116, 133, 135, 138, 139, 142, 145, 150, 153, 155, 159, 160, 163, 165, 169, 170, 175, 178, 187
Death ritual. *See* Ritual
Dehra Dun, 48, 112
Dharchula, 9, 10, 16, 44, 56, 57, 83, 89, 94, 96, 102, 103, 107–110, 113, 115, 118, 129, 138–143, 145, 149, 160, 176
Dogras, 12, 42–44, 46, 47
Doti, 44, 91, 99
  Raikas of Doti, 32
Dwelling, 1, 18, 112, 120, 131–133, 142, 159, 176, 178, 180, 187, 188

E
Ecology, 26
  new ecology, 128
Empire
  British, 7, 35, 48, 52, 58, 76, 77, 79, 185, 186
  Mongol (Tartary), 78
  Roman, 81, 105
  Russian, 48
Environmentality, 130, 155
Espionage, 103
Ethnicity, 15, 64, 84, 85, 114
Ethnogenesis, 78
Evil spirit (*syino*), 1, 156

F
First Opium War. *See* War
Fission, 29
Forest
  deforestation, 128
  Forest Act, 53
  Imperial Forest Department, 12, 53
  Joint Forest Management Program, 129
  montane forests, 10
  Pinus roxburghii-forests, 9
  Sal-forests, 9
  van panchayat, 129, 130, 134, 137, 138, 141, 142, 177, 178
Frontier, 6, 17, 25, 27, 29, 42, 48, 49, 51, 54, 55, 58, 63, 64, 94, 101, 186, 187
Fusion, 29

G
Gaddi, 159, 188
Galactic polity, 37, 38
Gandhi, 3, 100, 101, 106, 117
  Gandhian nationalism, 88, 104, 107
Garhwal, 8, 10, 24, 25, 31–34, 38, 61, 77, 82, 89, 92, 99, 107, 109–111, 113, 118
  Tehri Garhwal, 24
Gender, 4, 15, 36, 64, 87, 92, 93, 99, 116, 129, 148, 161
Gorkha, 38, 79
  Anglo-Nepalese war (*see* War)
Gosain, 38
Governmentality, 129, 130, 155
Great Game, 12, 48, 52
  Russian threat, 42
Great Trigonometrical Survey, 25, 48, 51–53
Gvan. *See* Ritual

H
Harijan, 3, 8
Heritage/heritagization, 77, 105, 114, 115, 119, 151, 154
High Asia, 6, 7, 17, 26, 63, 134, 186, 188
High Court, 109, 111, 112, 115
Himachal Pradesh, 44, 61, 89, 138, 159, 188
Hinduization, 15, 93, 96, 97
Humla, 32, 167, 175
Huniyas, 34, 109
Hydro-energy, 13, 136

I
Identity. *See* Tribal identity
Impression management, 26
India
  Five Year Plan, 83
  India's independence, 13, 14, 17, 32, 61, 87, 97, 100, 102, 116, 119, 136, 160, 187
  Indian Constitution, 81, 108
  Minority Sub-Committee, 76, 78, 79

Indo-Gangetic plain, 26, 78, 103, 128
Indo-Tibetan interface, 26, 188
Inner Line, 88, 142
Internationalism, 28

J
Jambu. *See* Medicinal plants
Jammu, 42, 45, 46
Johar valley, 8, 9, 32, 33, 41, 48, 56, 78, 91, 134, 138, 139, 141
Jongpen, 14, 55–59
Jumla, 32, 36, 37

K
Kashmir, 39, 48, 61
  Kashmiri traders, 41
Katyuri, 32, 34, 39, 91
KBPF. *See* Kumaon Bhotiya Peoples' Federation (KBPF)
Khadi, 100, 101
Kida. See Cordyceps sinensis
Knowledge, 16, 17, 35, 39, 40, 42, 52, 53, 76, 100, 128–132, 155, 165
Kumaon
  Kumaon and Uttarakhand Zamindari Abolition Land Reforms Act, 106
  Kumaon Bhotiya Peoples' Federation, 14, 15, 76, 77, 79, 80, 84, 87, 92, 95–100, 102, 104, 106, 116, 117, 137
  Kumaon kingdom, 32, 34
Kumbh melas, 38
Kutki. *See* Medicinal plants

L
Labe Dhura pass, 9
LAC. *See* Line of Actual Control (LAC)
Ladakh, 12, 33, 39, 44, 46, 48, 50, 61
  Ladakhi traders, 10, 41, 42, 44, 46
League of Nations, 28
Legal geographies, 28
Leprosy, 165, 175
Line of Actual Control (LAC), 62, 88
Lipu Lekh, 14, 46, 57, 58, 88, 141
Liquor (*cyakti*), 143
Livelihood, 4, 14, 77, 83, 128, 129, 133, 134, 178, 179, 188
  risk of livelihood failure (*see* Risk)
Livestock
  grazing, 178, 187
  horses, 56
  sheep and goats, 8, 9, 159
  yaks, 133, 159
Lok sabha, 115

M
MacMahon-Line, 62
Mana valley, 114
Mandal Commission, 107
Market, 4, 6, 12–15, 40, 45, 47, 58
  market-oriented border region, 13, 18, 128–180
Medicinal plants
  Aconitum heterophyllum (*atis*), 139
  Allium stracheyi (*jambu*), 139, 141
  Carum carvi (*thoya*), 139, 141
  Picrorhiza kurroa (*kutki*), 139
Miju, 2, 3
Milam, 33, 34, 48, 101, 134
Missionary(ies), 56, 57, 77, 94, 101
Mitra. *See* Trans-Himalayan trade
Mobility
  pastoral mobility, 6, 18, 128–180, 186, 188
  seasonal migration, 15, 106, 120, 133–142, 148, 155, 159, 175, 178, 179
  translocality, 188
Modernization, 84, 96
Munsiari, 10, 16, 83, 107–110, 115, 118, 129, 134, 138–141
Munsif, 56
Myth, 18, 30, 51, 77, 80, 132, 176, 178, 180, 187

N
Nanda Devi, 33–35, 37
Nepal, 3, 5, 7, 10, 14, 24, 27, 32, 36, 37, 49, 58, 59, 63, 99, 102, 186, 188
Niti valley, 40, 113
Non-governmental organization (NGOs), 138, 139, 141, 178
Non-regulation Province, 10, 17, 24, 42–48, 54, 63, 186
Non-state space. *See* Zomia
North-East Frontier Agency (NEFA), 61

O
Old Lady Jasuli, 87–92, 98–100, 104, 106, 116, 117
Oral history, 16, 29, 31
Other Backward Classes (OBCs), 107

P
Pahari, 8, 14, 32, 55, 98, 113, 150
Pakistan, 62, 85
Palti. *See* Buckwheat
Panch Shila, 61
Pandit. *See* Pundit
Pashmina. *See* Wool
Pastoralism, 102, 132
  tribal and transhumant strategy, 136, 187
Pasture
  bugyal, 10
  degradation, 128
  summer, 10, 158
  winter, 9, 142, 143, 158
Patwari, 42, 44, 46, 48, 55
Pilgrimage, 33, 38, 49, 51, 61, 89, 91, 113
Pithoragarh, 16, 103, 104, 108, 112, 139, 141
Political peshkar, 55, 59, 98
Princely state, 24, 28
Pundit, 25, 39, 49–53

Q
Quasi-sovereign polities. *See* Sovereignty

R
Railway, 12, 47
Rajput, 3, 32, 33, 109
Ralam, 8, 139
  Ralam pass, 9
Rambang, 93–100, 105, 114, 116, 117
Rang, 8, 97, 99, 104, 105, 115, 116, 119, 120, 150
Rangbang. *See Rambang*
Rawat, 33, 34, 37, 41, 48, 63
Reddish (*lapho*), 146
Resilience, 188
Risk
  of livelihood failure, 133, 178
  prevention, 178
Ritual
  alam, 145, 178
  cremation, 159
  death rituals, 171, 179
  gvan, 133, 172, 179
  ritualization, 142
  ritual journey, 160, 179
  sarat, 133, 160–162, 172, 176, 177, 179
  syimi thomo, 144, 147
Rule of law, 24, 54–60, 186
Russian threat. *See* Great Game

S
Sai, 35–37
  appellation, 33, 35
Salt, 3, 8, 10, 31, 36, 41, 42, 45, 47, 58, 60, 91, 145, 172
Sanskritization, 96
Sarat. *See* Ritual
Scale
  effect, 87, 92–107, 117, 187
  performativity of, 17, 87, 99, 118
  politics of, 86, 87
  scalar expression, 88, 107–116, 118, 187
  scalar politics, 17, 76–120, 187
Scheduled Caste, 8, 81–83, 106–108, 111
Scheduled Tribe (ST)
  adivasi, 82
  anusuchit janjati, 83
Shah dynasty, 34
Shauka, 8, 105, 116
Shawl-wool. *See* Wool
Shepherd, 9, 16, 129, 131, 132, 136–138, 142, 143, 147, 155, 157
Shinla pass, 9, 139
Sikh kingdom, 12, 42, 46
Sikh war. *See* War
Sino-Indian war. *See* War
Soul (*syimi*), 160, 164
South Asia, 27, 45, 81, 82, 92, 96, 105, 147, 151
Southeast Asia, 27, 63
Sovereignty
  de facto, 27
  de jure, 27
  imperial, 17, 24, 47, 53, 63, 187
  overlapping, 17, 27, 38, 54, 59, 63, 64, 186
  quasi-sovereign polities, 28
  regime, 27, 29
State policy, 5, 129
  transcultural rationality, 130
Supreme Court, 111, 112
Sustainability, 130, 155, 188
Syimi. *See* Soul
Syino. *See* Evil spirit

T
TAC. *See* Tribes Advisory Council (TAC)
Taskscape, 132, 176
Tatary. *See* Empire
Taxation, 7, 26, 55, 188

Territoriality, 25–42, 55, 186
confluent territories, 17, 27, 38, 54, 55, 59, 63, 64, 186
Thoya. *See* Medicinal plants
Tibet
Bod, 7
Ngari, 32
Tibetan plateau, 8–10, 23, 26, 36, 40–42, 47, 48, 51, 63, 88, 105, 186
Tibeto-Burman language family, 35, 36
rang-lvu, 35, 36
Tincal. *See* Borax)
Topology, 51
Tourism, 13, 15, 88, 89, 113, 136
Trans-Himalayan trade
mitra, 14, 58
sarji, 14, 58
Translocality. *See* Mobility
Transport, 14, 59, 102, 103, 143, 159, 179
Treaty of Amritsar, 12, 46
Treaty of Segauli, 24
Tribal
culture, 80
identity, 76–120, 187
status, 17, 87, 88, 113
Sub-Plan, 109, 136, 142
Tribe
Mongoloid border tribe, 12, 15, 79
Scheduled Tribe (*see* Scheduled Tribe (ST))
Tribes Advisory Council (TAC), 110
tribus, 81

U
Uttarakhand
independence, 112, 142, 187
movement, 118
Uttaranchal. *See* Uttarakhand
Uttar Pradesh (U.P.), 8, 13, 82, 88, 98, 103, 105–109, 118. *See*

V
Van panchayat. *See* Forest
Van Raji, 110

W
War
Anglo-Nepalese war, 42
Auckland's Folly, 43
First Opium War, 43
Sino-Indian war, 8, 15
war of maps, 62
Watershed, 61, 77
Watershed Development Program, 136
Wayfaring, 179
Weather, 18, 131, 133, 143, 145, 180, 187
Western Himalaya, 7, 44
Wife-giver, 154, 159, 161, 172–174, 179
Wife-taker, 154, 159, 161, 172–174, 177, 179
Woman
Bhotiya women, 92, 95
womanhood, 100, 101
Wool
mills, 12, 47, 102
pashmina, 10
shawl-wool, 12, 39, 44, 46, 47
trade, 12, 44, 46
woolen industries, 138
World Bank, 129, 136

Y
Yak. *See* Livestock
Yarsha gumbo. *See* Cordyceps sinensis

Z
Zomia
non-state space, 63
Zomia-thinking, 27, 63, 186, 188

MIX
Papier aus verantwortungsvollen Quellen
Paper from responsible sources
FSC® C105338

If you have any concerns about our products,
you can contact us on
ProductSafety@springernature.com

In case Publisher is established outside the EU,
the EU authorized representative is:
Springer Nature Customer Service Center GmbH
Europaplatz 3, 69115 Heidelberg, Germany

Printed by Libri Plureos GmbH
in Hamburg, Germany